Adsorptive Separation Process for Water and Gaseous Air Pollution Control

Adsorption is one of the efficient cost-effective separation processes that is in use for the removal of water and gaseous pollutants. This book provides a balanced mix of fundamental adsorption theory and application with a compilation of up-to-date developments and trends. It presents fundamental and advanced knowledge on solid/liquid and gas/solid separation unit operations and critically discusses the reaction-based mechanism of adsorption and adsorptive parameters estimation.

Key Features

Puts emphasis on adsorption for water and gas remediation, reviewing state-of-the-art of the technology as well as introducing beginners to adsorption

Explores factors affecting the adsorption separation process including types of adsorbents and composite nano-adsorbents

Includes batch and continuous unit operation and design calculations

Proposes development of a sustainable cost-effective water filter including arsenic-rich water purification

Discusses utilization of industrial and agricultural solid waste in alternative adsorbents during solid/liquid and gas/solid separation

This book is useful for researchers and graduate students in chemical and environmental engineering and analytical chemistry.

Adsorptive Separation Process for Water and Gaseous Air Pollution Control

Theory and Applications

Tushar Kanti Sen

CRC Press
Taylor & Francis Group
Boca Raton London New York

CRC Press is an imprint of the
Taylor & Francis Group, an **informa** business

Designed cover image: iStock Images

First edition published 2026
by CRC Press
2385 NW Executive Center Drive, Suite 320, Boca Raton FL 33431

and by CRC Press
4 Park Square, Milton Park, Abingdon, Oxon, OX14 4RN

CRC Press is an imprint of Taylor & Francis Group, LLC

ISBN: 9781032419770 (hbk)
ISBN: 9781032446554 (pbk)
ISBN: 9781003373223 (ebk)

DOI: 10.1201/9781003373223

Typeset in Palatino
by codeMantra

To my parents, late Jahar lal Sen and late Smt. Mira Sen

and to my family members and

to my all HDR (higher degree by research) students

Contents

About the author

Dr Tushar Kanti Sen is working as a Full Professor in Chemical Engineering of King Faisal University (KFU), Saudi-Arabia, since March 2022. Prior to KFU, he worked as an Associate Professor in Western Australia School of Mines (WASM): Minerals, Energy & Chemical Engineering of Curtin University, Perth, WA, Australia, during 2008 to March 2019 and is currently holding the same position as an Adjunct Associate Professor with Curtin University, Perth, Australia, since April 2019. Also, Dr Sen was a Chemical Engineering Discipline Coordinator and worked as a Senior Lecturer in School of Engineering of Edith Cowan University (ECU), Perth, Western Australia, Australia, from July 2019 to November 2021. Prior to joining Curtin, he worked at Universiti Teknologi PETRONAS (UTP), Malaysia, during 2006–2007 and worked with National Institute of Technology (NIT), Rourkela, India, during 1991–2005. Dr Sen obtained his B.Sc. (Hons) B.Tech. in Chemical Engineering from the University of Calcutta, India, and Master of Chemical Engineering from Jadavpur University, Kolkata, India. He obtained his Ph.D. in Chemical Engineering from Indian Institute of Technology (IIT) Bombay, Mumbai, India. He has over 30 years of teaching and research experience in Chemical Engineering & Environmental Engineering. Dr Sen is continuously engaged in undergraduate and postgraduate teaching and learning, supervision, research activities of the Department of Chemical Engineering. Dr Sen has an excellent track record in the broad field of Water, Energy & Environment and in the subfield research areas water and wastewater treatment process (chemical, physical, and biological treatment) including modelling & simulation, produced water Treatment, biomass-based biosurfactant in Chemical Enhanced Oil recovery (CEOR), adsorption of surfactants at solid/liquid interface, agricultural/industrial solid waste utilization in water treatment, flocculants/coagulating/chemical agents in dewatering, sustainable biomass-based adsorbent materials in water treatment, clay minerals/nanomaterials in water treatment, three-phase semi-fluidization operation in water treatment and rheology of sludge/crude oil emulsion and in separation processes, and process design. He has edited three research books on water and industrial wastewater treatment and clay minerals area. Dr Sen served as an Associate

Editor of a journal Adsorption Science & Technology during 2015–2020 and served as an Associate Editor of another journal Applied Water Science (AWSC) during 2019–2020. He has guest edited a Special Issue entitled Water Pollution Control Using Clay Minerals and Agricultural Biomass-Based Adsorbents published by Water journal and guest edited another special Issue entitled Solid/Liquid Interface Adsorption in Water and Wastewater Treatment published by Water (ISSN: 2073-4441). He has authored over 180 peer-reviewed international publications and a total google-scholar citations of above 15000 in his credits as on 10 October 2025. Few American patents are also in his credits. Dr Sen was leading externally industry-funded water engineering research group (more than 2 million dollars) responsible for research project execution. Dr Sen also supervised many higher degrees by research students (Ph.D./MS/M.Phil. /M.Tech.). He is a recipient of the HORN Scholarship-2014, 2012 Japan, "Ganga Ram Memorial Prize (Gold Medal)-2011, 2009" instituted by Institution of Engineers (India), "Certificate of Merit" in 2008 by the Institution of Engineers (India), "Shah-Schulman" award, instituted by IIChE, India, and an Australian Academy of Science scientific visit Award-2011 and winner of Gold Medal on his patent showcase event at IIFME 2025 held at Kuwait during 16–19 February 2025 for his outstanding contribution in research. He is also a recipient of "University Gold Medal" from Jadavpur University, Kolkata, India; "Visiting Fellowship" from ANU, Australia; "Visiting Professorship at the University of Hyogo, Himeji, Japan; Senior Research Fellowship" from MHRD, New Delhi, India; and Travel award from AICTE, New Delhi. He is also an associate with many professional activities such as a regular reviewer of many international journals, journal editor, journal editorial board members, session chair of international conferences, Ph.D. thesis external examinership from various universities, reviewer of research project proposal and also annual membership of many professional bodies including the IChemE, (U.K.), AIChE, International Adsorption Society, Australia Rheology Society, Int. Society for Colloid & Interface, Australian Association of Engineering Education, Indian Institute of Chemical Engineers, Institution of Engineers (India), Indian Institute of Public Health Engineers, Computer Society of India, Indian Institute of Metals, etc. Dr Sen gave many invited talks on adsorption, water and wastewater treatment processes, and air pollution in many universities and in many international conference platforms.

Preface

The availability of clean water and air remains a global challenge. Air and water pollution occurs when various contaminants (organic, inorganic, radioactive, etc.) are discharged into the environment without adequate treatment to remove these pollutants. The main sources of these toxic pollutants are industrialization, human activities such as agriculture and deforestation, unplanned urbanization, and other environmental and global changes. Major water pollutants include toxic metals, organic chemicals, pharmaceuticals, suspended solids/sediments, pathogens, nutrients, and salts. On the other hand, air pollutants primarily consist of gaseous pollutants such as carbon monoxide (CO), carbon dioxide (CO_2), nitrogen oxides (NO_x), sulphur oxides (SO_x), hydrogen sulphide (H_2S), and lead (Pb), along with particulate matter and odours. Today, carbon dioxide, a greenhouse gas, is the main pollutant warming the Earth. Its primary sources are power plants and human activities involving the burning of fossil fuels such as gasoline, natural gas, and coal. Polluted air and water not only cause material and vegetation damage but also have severe impacts on human health, including chronic respiratory diseases. Therefore, the sustainable and cost-effective development of removal techniques to address environmental pollution is a challenging and demanding task for engineers, scientists, and researchers. Furthermore, various federal and state government agencies, as well as environmental protection agencies, strictly enforce regulations to minimize pollutants from gaseous and liquid effluent streams. It is thus imperative to ensure all effluent streams are treated properly before being discharged into the environment or recycled.

Various physical, chemical, and biological separation processes are adopted as part of air and water pollution control strategies to achieve acceptable quality levels and comply with governmental and environmental regulations. Treatment technologies such as reverse osmosis, filtration, adsorption, absorption, chemical precipitation, coagulation, foam flotation, sedimentation, solvent extraction, electroplating, oxidation/reduction, ion exchange, activated sludge treatment, aerobic and anaerobic treatment, and membrane separation are employed to remove pollutants from gaseous and liquid effluents. Separation processes are generally used for both heterogeneous and homogeneous mixtures, with homogeneous mixture separation primarily relying on mass transfer principles, traditionally termed "mass transfer operations." These operations form a core module in chemical, process, mechanical, and environmental engineering curricula. Among various mass transfer separation techniques, adsorption stands out due to its simple design, universal applicability, ease of operation, high efficiency, cost-effectiveness, and potential for regeneration. Fundamentally, adsorption is a process where selective concentration or retention of one or more components of a mixture occurs on a solid surface or at a solid-fluid interface. Hence, it is an interfacial separation technique involving solid/liquid or solid/gas systems. The solid that adsorbs the pollutant is called the adsorbent, and the accumulated pollutant is termed the adsorbate.

Given its industrial importance and role as a major unit operation in chemical and allied process industries—including reaction purification, gas and liquid processing, and catalysis—adsorption is integral to separation processes. Considering the importance of air and water quality and environmental concerns, it is worthwhile to address the state-of-the-art in adsorption for removing gaseous and water pollutants. The nature, types, and characteristics of adsorbents, along with various physicochemical parameters, are the main factors controlling the effectiveness of the adsorptive separation process. Additionally, mass transfer and separation processes constitute a core component of undergraduate and postgraduate chemical, mechanical, and environmental engineering courses. Adsorptive separation is a key mass transfer operation, and contemporary course structures require a balanced mix of fundamental concepts and applications, along with up-to-date developments and trends. With these objectives, this textbook was written to provide a comprehensive understanding of the fundamentals, advancements, and applications of adsorptive separation processes for reducing potential pollutants (inorganic/organic) in gaseous and liquid effluents. The book also compiles extensive literature on the adsorptive removal of water and air pollutants, which will benefit engineers, scientists, and researchers. The key features of this research-based textbook are as follows:

- Covers the fundamentals and applications of adsorption processes, comparing them with other treatment methods.
- Presents state-of-the-art applications of solid/liquid and gas/solid adsorptive separation processes.
- Explores factors affecting adsorption processes, including adsorbent types and surface characteristics.
- Discusses adsorption theory, mechanisms, and model and design parameter calculations.
- Examines adsorption kinetics and equilibrium models for batch and continuous processes, highlighting common errors in recent publications.
- Discusses the surface reaction-based Surface Complexation Model (SCM) for adsorption mechanisms.
- Includes theory and design calculations for continuous adsorption operations, such as fixed-bed, fluidized-bed, and semi-fluidized bed systems.

Although this manuscript has been thoroughly checked, errors may still remain. I would greatly appreciate it if readers would kindly bring any mistakes to my attention. This book is intended for professionals, practising engineers, scientists, researchers, academics, and undergraduate and postgraduate students interested in this specific area.

I take this opportunity to express my sincere gratitude to Dr. Suleiman Mousa, my youngest beloved colleague and friend at KFU, Saudi Arabia, for his invaluable contributions to partial peer-reviewing, rewriting mathematical equations, improving some figures qualities and editing the English grammar in all chapters. I also wish to thank all my present colleagues at KFU, Saudi Arabia, and my past colleagues at Curtin University and ECU in Perth, Australia, for their continuous support and encouragement in writing this book. My heartfelt thanks go to my past Ph.D. and M.Phil. students, especially Dr. Sharmeen Afroze, Dr. Sara Dawood, and Dr. Subrata Biswas of IIT Roorkee, India, for allowing me to reuse plots and literature from their Ph.D. theses in this book. I extend special thanks to my dear friend and colleague, Prof. Bankim Chandra Ray of NIT Rourkela, India, and Associate Professor Dr. Chi Phan of Curtin University, Australia, for their invaluable suggestions and for reviewing my initial book proposal. I am also deeply grateful to my family—my brothers, my only sister, Tumpa, my sisters-in-law, especially Associate Professor Sujata Sen and Chemical Engineer Sarbari Saha—for their support and best wishes.

Finally, I would like to acknowledge Dr. Gagandeep Singh of CRC Press, Taylor & Francis Group, for his invitation to write this book and for promptly addressing all my queries. I also thank the CRC Press staff for their assistance with the editorial process. Lastly, my sincere thanks to my wife, Kabita, and my son, Anindya, for their patience and support throughout this endeavour.

Dr Tushar Kanti Sen
Professor
Chemical Engineering Department
College of Engineering
King Faisal University
P.O. Box: 380, Al-Ahsa 31982
Kingdom of Saudi-Arabia

1

Fundamentals on Adsorption, Significance and Its Application in Water and Air Pollutants Removal

1.1 Introduction

The sustainable and cost-effective remediation of water and air pollutants to produce clean water and air is a significant challenge for scientists, researchers, and engineers worldwide. Pollution occurs when various contaminants—organic, inorganic, radioactive, and others—are released into the environment without adequate treatment. The primary sources of these toxic pollutants include industrialization, human activities such as agriculture and deforestation, unplanned urbanization, and broader environmental and global changes. Major water pollutants encompass toxic metals, organic chemicals, pharmaceuticals, suspended solids, pathogens, nutrients, and salts, among others. Conversely, air pollutants predominantly consist of gases such as carbon monoxide (CO), carbon dioxide (CO_2), nitrogen oxides (NO_X), sulphur oxides (SO_x), hydrogen sulphide (H_2S), lead (Pb), particulate matter, and odours. Therefore, removing these contaminants through appropriate treatment methods before environmental release is essential. Among available techniques, adsorption has emerged as a highly effective and widely used separation method for water and air pollution control (Sen, 2023). This textbook section will discuss the fundamentals of the adsorption process, its applications, and the various factors influencing adsorption in the removal of water and air pollutants.

1.2 Adsorption, Adsorbate, Adsorbent, Absorption, and Desorption

Adsorption is defined as the transfer of one or more solute molecules from a bulk liquid or gas phase to the surface of a solid adsorbent, where they

DOI: 10.1201/9781003373223-1

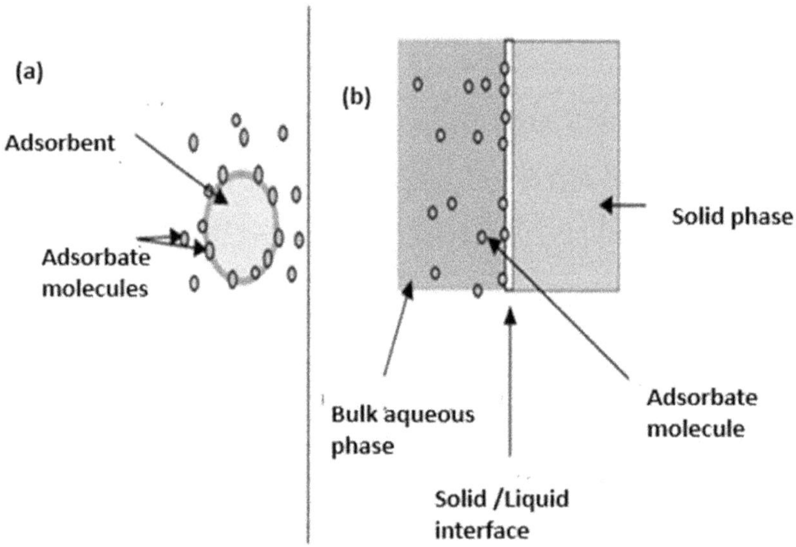

FIGURE 1.1
Adsorption at the solid/liquid interface (Reproduced from Dawood, 2018; Dawood, Sen et al., 2017a, with permission from Nova Science Publishers, Inc. [https://novapublishers.com/permission-to-reuse-content]). Courtesy of Slejko, 1985

are retained (see Figure 1.1). The solid that adsorbs the solute component is termed the *adsorbent*, while the adsorbed solute is known as the *adsorbate* (Figure 1.1a). This process occurs at the solid/liquid or gas/solid interface, as illustrated in Figure 1.2 for a gas-phase bulk fluid (Geankoplis, 2003; Seader & Henley, 1998). When a bulk solution containing adsorbable solutes contacts a highly porous solid surface, diffusional mass transport and intermolecular forces attract solute molecules from the solution phase, concentrating them on the solid adsorbent surface. Common commercial adsorbents include activated carbon, zeolites (molecular sieves), silica gel, and activated alumina (Dutta, 2007).

Adsorption is an advanced separation process wherein solute molecules from a bulk fluid phase bind to the surface of a solid porous material (Figures 1.1 and 1.2). This process results from interactions between the adsorbate molecules and the adsorbent's solid surface (Dutta, 2007). However, the specific transfer of solute species from the homogeneous mixture to the surface or into the bulk of a solid (or a liquid) is called **Sorption (**Dutta, 2007**).** Sorption, meanwhile, describes the transfer of solute species either to a surface or into the bulk of a solid or liquid. Sorption generally encompasses both adsorption and absorption. When solute species concentrate near the solid surface, the process is termed adsorption, whereas absorption involves the bulk distribution of solute molecules throughout the liquid phase (Dutta, 2007; Tien, 1994).

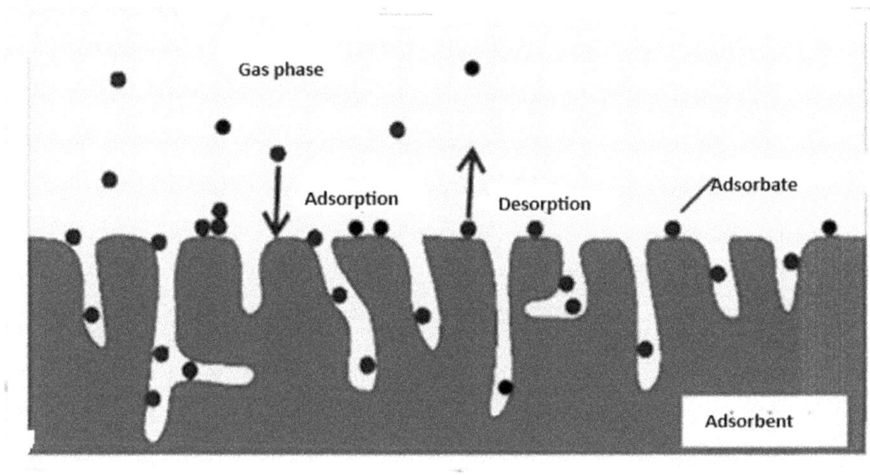

FIGURE 1.2
Adsorption at the gas/liquid interface

In gas absorption, a separation occurs where gaseous solute species are isolated from a gas mixture via contact with a liquid solvent. The reverse process, known as *stripping* or *desorption,* removes bound adsorbate molecules from the solid adsorbent surface (Figure 1.2). Desorption is crucial for understanding adsorption mechanisms and regenerating the adsorbent. It also allows for the recovery of valuable products from the loaded adsorbent. Both adsorption and absorption are categorized as diffusional equilibrium separation processes (Tien, 1994). In a continuous separation process, gas absorption via a packed bed column operates in three phases (gas, liquid, and solid packing material) in steady state, while packed bed adsorption operates as a two-phase (gas and solid or liquid and solid) in non-steady-state process.

1.3 Classification and Mechanism of Adsorption and Its Application

Adsorption on a solid surface occurs due to the surface's "affinity" for specific components. When adsorption results from weak Van der Waals or short-range forces, it is known as *physical adsorption*. In contrast, *chemical adsorption* involves covalent or ionic bond formation between the adsorbate and adsorbent through electronic transfer, making it irreversible in nature (Dawood, 2018). Physical adsorption can lead to monolayer or multilayer formation on the solid surface and is nonspecific, whereas chemical adsorption is highly specific and only forms a monolayer. The heat of chemisorption is

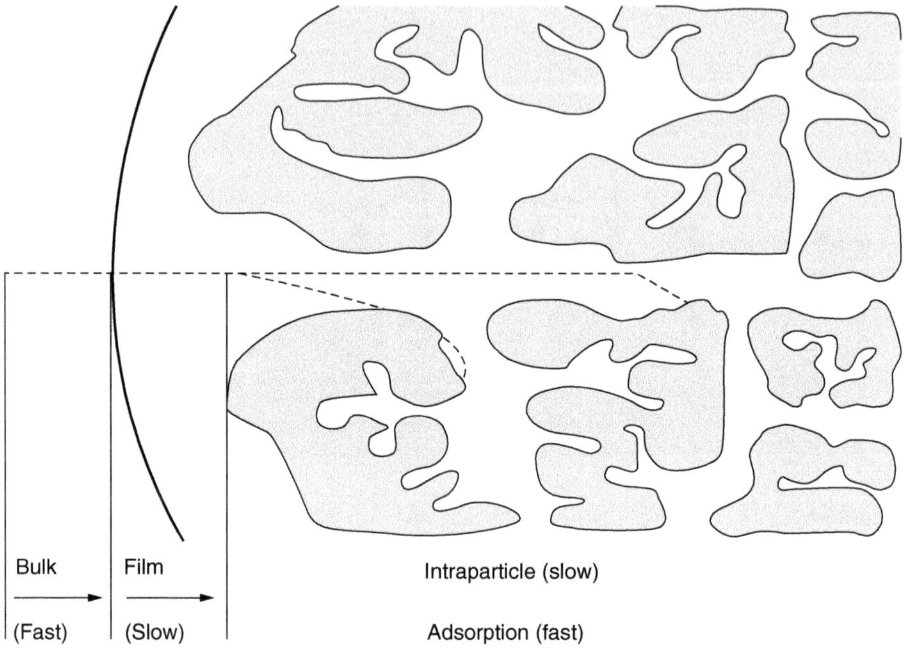

FIGURE 1.3
Mechanistic diffusional mass transport of adsorbate solute species from the bulk fluid phase to the solid adsorbent material (Courtesy of Seader and Henley, 1998)

greater than that of physical adsorption. Fundamentally, physical adsorption is an equilibrium diffusional process, involving the following four steps (see Figure 1.3):

- Bulk fluid transport
- Film transport
- Intra-particle (pore and/or surface) diffusion
- Physical attachment

Liang et al. (2013) proposed a surface adsorption mechanism for aqueous-phase Cr^{6+} onto lignin-based resin (LBR) via physical adsorption, as shown in Figure 1.4. The adsorbent surface contains functional groups—such as alcohol, ketonic, and carboxyl groups—that facilitate Cr^{3+} adsorption at low solution pH. The surface charge of the adsorbent varies with solution pH, causing adsorbed Cr^{6+} anions to transform into Cr^{3+} due to protons on the surface (El-Baz et al., 2020).

In chemical adsorption, the predominant mechanism depends on the activation energy of the reaction, resulting in surface complex formation between adsorbate and surface functional groups of the adsorbent, often through covalent bonding and ion exchange mechanisms, as shown in Figure 1.5. For

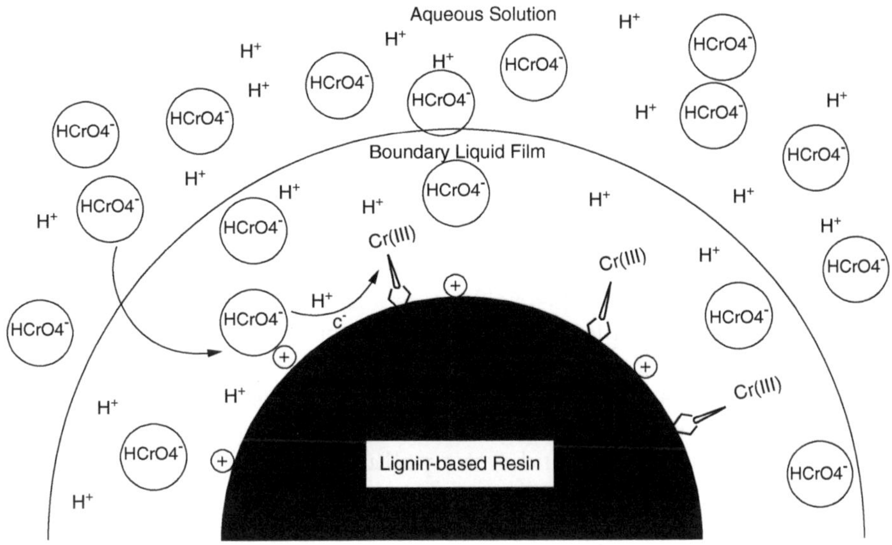

FIGURE 1.4
Physical adsorption of Cr^{6+} by surface accumulation

Source: Liang et al., 2013; El-Baz et al., 2020, with written permission.

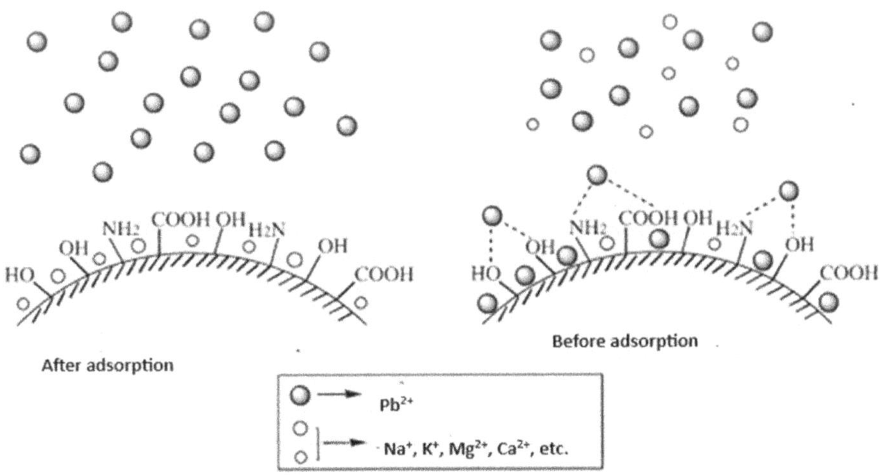

FIGURE 1.5
Pb^{2+} adsorption on FCCL adsorbent through ion exchange mechanism in chemical adsorption

Source: El-Baz et al., 2020; Chen et al., 2010, with written permission.

instance, the adsorption of Pb^{2+} on fallen Cinnamomum camphora leaves (FCCL) involves chemisorption through chelation and surface complex formation, with ion exchange of Na^+, Ca^{2+}, K^+, and Mg^{2+}. Chemisorption arises from electrostatic attraction between the charged adsorbent surface and adsorbate ions, forming covalent bonds (Dutta, 2007; El-Baz et al., 2020).

As a separation process, adsorption is widely applied in purification and other commercial applications, commonly using commercial adsorbents. For certain purification challenges, such as water and air purification, adsorption presents a highly effective solution. Consequently, adsorption is beneficial for both purification and industrial bulk separation. Table 1.1 provides a partial list of adsorption applications with commercial adsorbents like activated carbon, zeolite, silica gel, and activated alumina.

Given the importance of air and water quality and environmental considerations, it is essential to address the state-of-the-art applications of adsorption for removing gaseous and water pollutants. Therefore, this textbook focuses specifically on such water treatment and air pollutant removal applications. The nature and type of adsorbents, along with various physicochemical parameters, are critical for the effectiveness of the adsorption

TABLE 1.1

Industrial applications of adsorptive separation using various commercial adsorbent materials (Dutta, 2007; Tien, 1994)

Adsorption system	Application/uses
Liquid/solid phase adsorption	• Decolorizing and drying petroleum fractions • Decolourizing crude sugar syrups and vegetable oils • Widely used in water filtration processes • Purification of wastewater and industrial waste as part of physicochemical or tertiary treatment • Removal of organic and inorganic pollutants from water • Odour and taste removal • Sulphur compound removal • Bulk separation of normal paraffins/iso-paraffins, BTX • Separation of H_2O/organics, oxygenated and halogenated organics • Fructose/glucose bulk separation
Gas/solid phase adsorption	• Air pollutant removal • Dehydration of gases • Removal of CO_2 and sulphur from natural gas • Separation of normal paraffin from iso-paraffin aromatics • Removal of organics, odours, and toxic agents from gas/air • Nitrogen production from air • Drying of air and other gases • Hydrocarbon capture from emissions • SO_2/vent stream purification • Indoor air pollutant removal • O_2/N_2 bulk gas separation • Gas purification for H_2O/cracker gas, natural gas, and synthesis gas

process. This book aims to provide a comprehensive understanding of the fundamentals, advancements, and applications of adsorptive separation in reducing potential pollutants (inorganics/organics) from gaseous and liquid effluents.

1.4 Existing Separation Treatment Technologies and the Superiority of Adsorption in Water and Wastewater Treatment

Protecting water resources from contamination and treating pollutant-laden effluents efficiently and cost-effectively is a critical challenge for engineers, scientists, and researchers. Numerous conventional and advanced treatment strategies have been developed to remove inorganic and organic contaminants from water and wastewater, achieving acceptable water quality standards that comply with governmental and environmental regulations (Sen, 2023, 2015). These strategies, typically classified as physical, chemical, and biological processes, are effective for water treatment in various ways, as shown in Figure 1.6. Common treatment methods include membrane separation, reverse osmosis (RO), filtration, adsorption, chemical precipitation, coagulation, foam flotation, solvent extraction, electroplating, evaporation, oxidation/reduction, ion exchange, activated sludge, aerobic and anaerobic treatments, electrolysis, magnetic separation, and desalination, each with varying levels of success (Abdolali et al., 2016; Afroza & Sen,

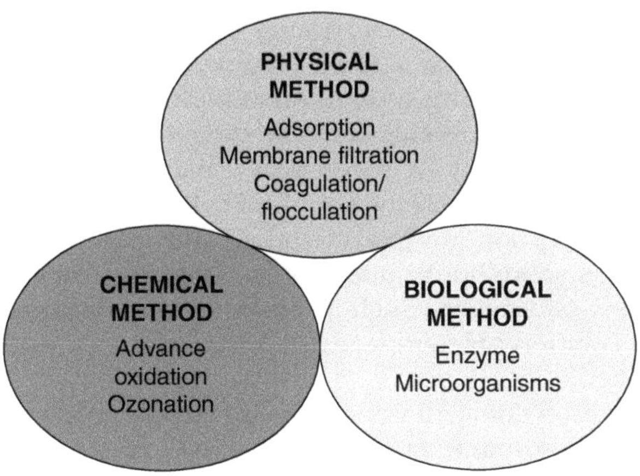

FIGURE 1.6
Traditional and advanced treatment methods in water and wastewater treatment

Source: Rashid et al., 2021, with written permission.

2018; Barakat & Schmidt, 2010; Biswas et al., 2019; Crini, 2006). However, many traditional technologies have notable drawbacks, including high power consumption, significant maintenance and operating costs, regeneration issues, and reduced efficiency over time. To address these limitations, advanced technologies such as membrane-based processes, photocatalysis, advanced oxidation, and adsorption-based technologies have been developed for water and wastewater treatment. Among these, adsorption-based separation is considered one of the most effective and widely used techniques for removing various pollutants from water and wastewater due to its simple operation, design simplicity, high separation efficiency (especially at low pollutant concentrations), molecular-level selectivity, low energy consumption, ability to remove multiple contaminants, and minimal secondary pollution. These advantages make adsorption a cost-effective and sustainable option (Ghosh et al., 2023; Liu et al., 2020; Sen, 2023).

Industrial and agricultural waste are primary sources of inorganic pollutants, which enter water bodies through human activities such as mining, mine drainage, smelting, metallurgical processes, and various chemical industries. Heavy metal pollutants are particularly concerning due to their acute toxicity, even at low concentrations in water. Thus, the adsorptive removal of various pollutants—such as heavy metals, radionuclides, anions, and organic dyes—has become essential and is increasingly employed (Sen, 2023). Similarly, various gaseous pollutants, including CO_2, SO_2, and particulates, are effectively removed through gas/solid or three-phase gas-solid-liquid adsorption operations.

Adsorption operations can be conducted in either batch or continuous (dynamic) modes, especially in large-scale treatment systems. In continuous mode, common setups include fixed bed or packed bed adsorption columns, fluidized beds, semi-fluidized beds, and bubble columns. Batch adsorption, a small-scale operation, occurs in a closed system where a fixed amount of adsorbent contacts a fixed volume of adsorbate solution, typically with an efficient stirring system. Adsorbent capacity is a crucial factor in selecting effective adsorbents for removing aqueous-phase pollutants, determined experimentally or theoretically using isotherms and kinetic models. Chapter 3 will discuss the details of batch kinetics and equilibrium experimental adsorption design methods and parameters calculation. Adsorption at the solid/liquid interface is influenced by various physicochemical process parameters, such as adsorbent characteristics, adsorbate concentration, solution pH, temperature, adsorbent dose, and the presence of salts or impurities.

The subsequent section will review the adsorption capacity and discuss the influence of these parameters on the solid/liquid adsorption process. Additionally, it will compile batch adsorption results for heavy metal ions and organics under various physicochemical process conditions, while Chapter 3 will also focus on continuous adsorption column operations, including fixed bed, fluidized bed, and semi-fluidized bed unit operation systems.

1.5 Factors Affecting the Adsorptive Separation Process

Various commercial adsorbents, such as activated carbon, zeolites, molecular sieves, silica gel, and activated alumina, are widely used for the adsorptive removal of inorganic and organic contaminants from aqueous phases. Despite the effectiveness of these materials, the high cost of commercial adsorbents has motivated the development of alternative, full-scale adsorption processes using natural or treated solid wastes and other low-cost synthetic adsorbents. Coal-based activated carbon, known for its high specific surface area, porous structure, and functional groups (e.g., carboxyl, carbonyl, phenolic, amine), is highly effective at binding metals to form surface complexes (Deliyanni et al., 2015; Fu & Wang, 2011). However, due to its cost and limited regeneration capacity, research is increasingly focused on alternative solid waste-based adsorbents. These alternatives include agricultural waste, aquatic and terrestrial biomass, local waste biomass, zeolites, clay materials, industrial solid waste, fullerene, and carbon nanotubes (Burakov et al., 2018). For an in-depth exploration of these alternatives, readers may refer to recent review articles on solid waste-based inorganic and organic adsorbents for heavy metal ion removal (Afroza & Sen, 2018; Ahmad et al., 2017; Chen et al., 2022; Gisi et al., 2016; Ogunlalu et al., 2021; Sen, 2023; Tee et al., 2022).

Numerous physicochemical factors influence the adsorption process and the maximum adsorption capacity of adsorbents for removing inorganic heavy metal ions and organic dyes from water and wastewater. Key influencing factors include adsorbate-adsorbent interactions, adsorbent surface area and pore structure, surface chemistry, adsorbate properties, ion effects, particle size, pH, temperature, and contact time (Afroza & Sen, 2018; Asadi et al., 2024; Crini, 2006; Sen, 2023). The initial concentration of adsorbate, solution pH, adsorbent dose, temperature, contact time, presence of ions, and stirring speed are also significant parameters affecting the adsorption process, as discussed below (Dawood, 2018). Understanding these physicochemical factors is crucial for optimizing and scaling up adsorption processes for industrial applications. The following factors specifically affect adsorption:

- **Surface Area of Adsorbent**: Larger surface areas correspond to greater adsorption capacities.
- **Particle Size of Adsorbent**: Smaller particle sizes reduce internal diffusion and mass transfer limitations, facilitating equilibrium and enhancing adsorption efficiency.
- **Contact Time or Residence Time**: Longer contact times generally lead to more complete adsorption, although limited by the adsorbent's capacity.
- **Solubility of Solute (Adsorbate) in Liquid (e.g., Wastewater)**: Materials with low solubility in water are easier to remove than those with high solubility.

- **Affinity of Solute for the Adsorbent**: The slightly polar surface of activated carbon more readily adsorbs non-polar materials than polar ones.
- **Molecular Size Relative to Pore Size**: Adsorption is influenced by the compatibility of molecule size with pore size.
- **Degree of Ionization of Adsorbate Molecules**: Neutral molecules are more readily adsorbed than highly ionized species.
- **pH**: The degree of ionization, and thus the adsorption effectiveness, is impacted by the pH of the solution.

1.5.1 Effect of Solution pH on Solid/Liquid Adsorption Process

Solution pH plays a significant role in altering adsorbent surface charges, the degree of ionization, and adsorbate speciation, thereby affecting adsorption capacity (Sen, 2023). Specifically, pH influences the surface charge of adsorbents and the ionization of adsorbates, which in turn impacts dye uptake (Asadi et al., 2024). Numerous studies report experimental and modelling data on cationic adsorption, particularly for heavy metal ions at mineral/aqueous interfaces, across different pH values, metal ion concentrations, solid-liquid ratios, and temperatures (Angove et al., 1998; Dzombak & Morel, 1986; Lothenbach et al., 1998; Neubaker et al., 2000; Sen, 2001; Schroth & Sposito, 1998; Trivedi & Axe, 2000; Vulava et al., 2000).

Initial pH affects both site dissociation and adsorbate solution chemistry (e.g., hydrolysis, complexation with organic/inorganic ligands, redox reactions, precipitation) (Afreza, 2016a). The adsorption process heavily depends on initial solution pH and the adsorbent's point of zero surface charge ($pH_{P}zc$), which defines the adsorbent's electrokinetic properties (Afroza, 2016a; Dawood, 2018). Using the solid addition method, $pH_{P}zc$ can be determined; cationic adsorbate adsorption is favoured at pH > $pH_{P}zc$ (indicating a negatively charged surface), while anionic adsorption is favoured at pH < $pH_{P}zc$ due to proton presence (Salleh et al., 2011). Additionally, H^+ and OH^- ions determine potential differences in adsorbate-adsorbent interactions, affecting electrostatic attraction or repulsion.

For cationic adsorbates, adsorption efficiency increases with pH, as lower pH causes positive surface charge and competition with H^+ ions, whereas higher pH increases OH^- availability, enhancing adsorption. Conversely, for anionic adsorbates, lower pH favours adsorption. High pH may cause metal hydroxide precipitation, so pH should be maintained just above $pH_{P}zc$ to avoid this during heavy metal ions adsorption (Biswas et al., 2019a, 2019b, 2019c).

McKenzie (1980) observed that cation adsorption on goethite and hematite follows a sigmoidal curve, with adsorption increasing over a narrow pH range (the "adsorption edge"). The affinity of metals for oxide surface OH groups align with their hydrolysis constants; for goethite, the adsorption strength follows $Cu^{2+} > Pb^{2+} > Zn^{2+} > Cd^{2+} > Ni^{2+} > Mn^{2+}$, while on hematite, Pb and Cu positions are reversed (McKenzie, 1980; Sen, 2001). Ferrihydrite follows a different order: $Cr^{3+} > Pb^{2+} > Cu^{2+} > Cd^{2+} > Zn^{2+} > Ni^{2+} > Ca^{2+}$ (Dzombak & Morel, 1990). Metal adsorption on oxides involves two types of active sites—low and high energy—favouring high-energy sites initially (Benjamin & Leckie, 1981).

Several studies have examined pH effects on heavy metal ion adsorption, including work by Tamura and Furuichi (1997) on MnO_2 and Fe_2O_3, showing increased adsorption density with pH. Qing-du et al. (1997) found copper adsorption on illite surfaces significantly depends on pH, with uptake increasing from pH 4.5. Angove et al. (1998) showed that Cd^{2+} and Co^{2+} adsorption on kaolinite occurs at low pH through ion exchange at sparse, permanently charged sites, while at high pH, adsorption occurs via inner-sphere complexation.

Experimental studies, such as those by Afroze et al. (2016a, 2016b) on methylene blue (MB) adsorption onto eucalyptus bark, have identified zero points of charge (pH_{PZC}) that maximize adsorption capacity at specific pH levels. Kolodyńska et al. (2017a) reported that Cu^{2+}, Zn^{2+}, and Cd^{2+} adsorption by commercial activated carbon peaks near pH 5, while Mallampati et al. (2015) found increasing pH enhances Pb^{2+} and Ni^{2+} adsorption but decreases $Cr_2O_7^{2-}$ uptake. Ahmad and Danish (2022) mentioned in their review article that Mallampati et al. (2015) reported the results of solution pH effect on the adsorptive removal of aqueous phase Pb^{2+}, Ni^{2+} and $Cr_2O_7^{2-}$ by avocado peel adsorbent. They found that the percentage removal of cationic Pb^{2+} and Ni^{2+} adsorption was increased with the increase in solution pH, whereas the adsorption of anionic $Cr_2O_7^{2-}$ was decreased with the same increasing solution pH. Dawood et al. (2017b) observed similar trends with Ni^{2+} adsorption on pinecone biochar.

Conversely, studies have shown different optimal pH levels for organic adsorbates. For example, Yildiz (2024) reported that MB adsorption on forest waste-based activated carbon increased with pH, while Khalid et al. (2015) found that Acid Blue 25 adsorption on rubber leaf powder peaked at pH 2. Pavan et al. (2014) showed that Crystal Violet adsorption by papaya seed powder is highest at pH 8. Zazycki et al. (2018) found that Reactive Red 141 dye removal by pecan nutshell biochar peaked at pH 2, as shown in Figure 1.7.

Table 1.2 provides a summary of various batch adsorption studies, detailing the maximum adsorption capacities and removal efficiencies of dyes and heavy metals by different adsorbents across varying solution pH levels.

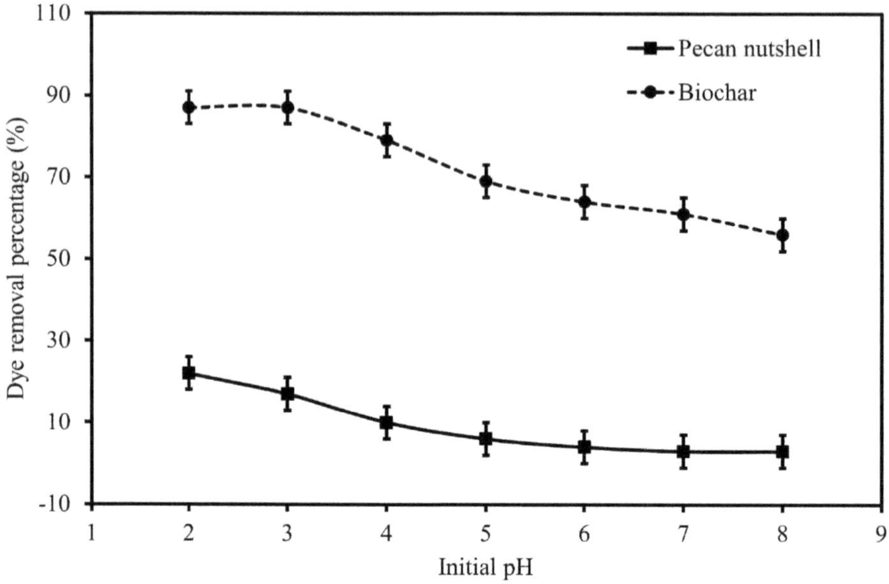

FIGURE 1.7
Effect of solution pH on the removal of Reactive Red 141 dye by raw pecan nutshell and biochar

Source: Zazycki et al., 2018, with written permission.

TABLE 1.2

Compilation of batch adsorption results for the removal of dyes and heavy metal ions at varying solution pH (Afroza & Sen, 2018; Dawood, 2018)

Adsorbent	Adsorbate	pH range	Removal range	Reference
Raw pecan nutshell	Reactive Red 141	2–8	85–60%	Zazycki et al. (2018)
Pecan nutshell char	Reactive Red 141	2–8	23–5%	Zazycki et al. (2018)
Korean cabbage biochar	Crystal Violet	4–11	Increase	Sewu et al. (2017b)
Bentonite	Al (III)	2–9	52–96%	Chai et al. (2017)
Kaolin	Al (III)	2–9	40–95%	Chai et al. (2017)
Commercial activated carbon	Cu (II)	5	80.6%	Kołodyńska et al. (2017b)
Fe_3O_4/kaolin composite	Direct Red	3–11	90–63%	Magdy et al. (2017)
Mushroom biochar	Crystal Violet	2–8	Increase	Sewu et al. (2017a)
Banana peel biochar	Pb (II)	4–7	70–90%	Zhou et al. (2017)
Mango leaf	Methylene Blue	2.5–10	10–40 mg/g	Uddin et al. (2017)
Chitosan/carbon composite	Methylene Blue	3–11	80–93 mg/g	Marrakchi et al. (2017)
Chickpea shell	Acid Blue 25	2–12	83–20%	Krishna et al. (2017)

(Continued)

TABLE 1.2
(Continued)

Adsorbent	Adsorbate	pH range	Removal range	Reference
Black-eyed biomass	Pb (II)	2–6	35–93%	Guyo and Moyo (2017)
Tea leaf activated carbon	Orange G	2.5–11	198–85 mg/g	Goswami and Phukan (2017)
Bamboo leaf	Methylene Blue	2–12	8–84.7%	Zhu et al. (2016)
Cucumis peel	Acid Blue 113	2–11	97.7–75%	Lee et al. (2016b)
Eucalyptus bark (EB)	Zn (II)	3–5.1	20–40%	Afroza et al. (2016c)
Base-modified eucalyptus bark	Zn (II)	3–5.1	40–69%	Afroza et al. (2016c)
Eucalyptus bark	Methylene Blue	2.5–10	58–89%	Afroza et al. (2016b)
Iron-manganese oxide/kaolin composite	Crystal Violet	4–11	27–94.8%	Khan et al. (2015)
Rapeseed oil cake activated carbon	Ni (II)	1–6	30–100 mg/g	Uçar et al. (2015)
Rapeseed oil cake activated carbon	Pb (II)	1–6	25–110 mg/g	Uçar et al. (2015)
Pinecone activated carbon	Congo Red	2.6–8.2	77–7.6%	Dawood et al. (2014a, 2014b)
Coal fly ash	Ni (II)	3–10	5–95%	Sočo and Kalembkiewicz (2013)
Coal fly ash	Cu (II)	3–10	2–98%	Sočo and Kalembkiewicz (2013)
Zeolite	Methylene Blue	5–12	99–77%	Rida et al. (2013a)
Raw pinecone	Congo Red	3.6–11	5.8–60.5%	Dawood and Sen (2012)
Acid-modified pine cone	Congo Red	3.6–7.5	16–79%	Dawood and Sen (2012)
Pine tree leaves	Methylene Blue	2–9	20–80%	Yagub et al. (2012)
Kaolin	Zn (II)	3.4–8	6–62.5 mg/g	Aries and Sen (2009)
Forest waste-based activated carbon	MB Dye	2–8	55–79 mg/g	Yildiz (2024)

1.5.2 Effect of Initial Adsorbate Concentration and Contact Time on Solid/Liquid Adsorption

The adsorptive removal of inorganic and organic solutes is crucial for developing sorbent-based water treatment technologies, particularly for pollutants like dyes and heavy metals. The initial concentration of adsorbates affects the relationship between the concentration of adsorbates and the availability of active sites on the adsorbent surface. This effect can be studied by preparing adsorbent–adsorbate solutions with a fixed adsorbent dose, fixed solution pH and temperature and varying initial concentrations of adsorbate, shaking the mixture until equilibrium is reached (Afroza & Sen, 2018; Salleh et al., 2011; Yagub et al., 2014).

High initial solute concentrations increase competition for available adsorption sites, potentially lowering overall removal efficiency (Afroze et al., 2016b; Biswas et al., 2020a, 2019b; Dawood et al., 2017b; Kumari et al., 2020). As the initial adsorbate concentration rises, the percentage removal tends to decrease, while the amount of dye or heavy metal ions adsorbed per gram of adsorbent increases (Dawood, 2018). This initial concentration provides a driving force to overcome mass transfer resistance, facilitating adsorption but eventually reaching an equilibrium stage when the adsorbent's active sites are saturated (Uddin et al., 2017). The adsorbed amount per unit mass, q_t (mg/g), increases with higher initial adsorbate concentration due to decreased resistance in the bulk solution (Foo & Hameed, 2011; Purkait et al., 2007; Sen et al., 2011).

Ding et al. (2016) observed that the equilibrium adsorption, q_e (mg/g), of Cd (II), Cu (II), Zn (II), and Ni (II) ions by Hickory wood biochar and its modified form increased with initial metal ion concentration from 2 to 100 mg/L (Sen, 2023). Afroze et al. (2016b) studied MB dye adsorption on raw eucalyptus bark and reported a decrease in percentage dye removal from 85.9 to 38.8% as initial dye concentration increased from 20 to 70 mg/L, while the amount adsorbed increased from 43 to 67.9 mg/g over the same concentration range. Similarly, Said et al. (2017) investigated diazonium blue dye

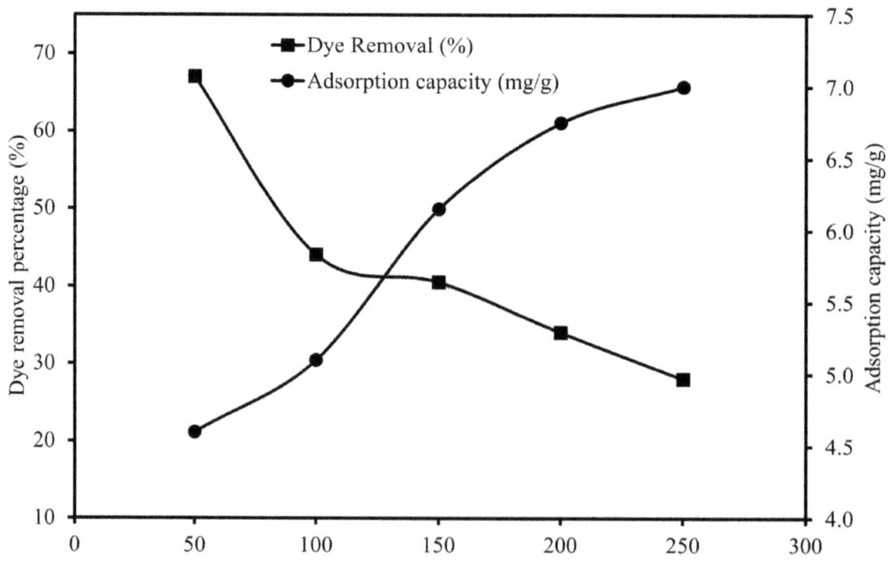

FIGURE 1.8
Effect of initial diazonium blue dye concentration on percentage dye removal and adsorption capacity by modified sugarcane bagasse. All other parameters were held constant (Said et al., 2017)

removal using acid-modified sugarcane bagasse and found that dye removal decreased from 66.4 to 28%, while adsorption capacity increased from 4.5 to 7 mg/g with initial dye concentrations between 50 and 250 mg/L (Figure 1.8). Kilic et al. (2013) demonstrated a three-stage process for Ni^{2+} adsorption on almond shell biochar, consisting of an initial rapid rate, a slower rate, and eventual equilibrium at around 240 minutes (Figure 1.9) (Sen, 2023).

Recent studies on the removal of dyes and heavy metals at various initial adsorbate concentrations are summarized in Table 1.3. Readers are encouraged to refer to comprehensive review articles by Sen (2023) and his research group (Afreza & Sen, 2018) for further insights and an extensive compilation of updated literature.

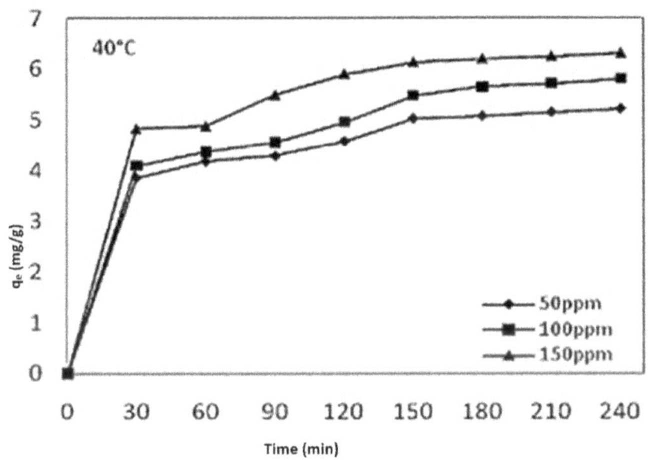

FIGURE 1.9
Effect of contact time on Ni (II) ion adsorption capacity by almond shell biochar. Adsorbent dose = 7 g/L, solution pH = 7, and T = 40°C (Kilic et al., 2013)

TABLE 1.3

Removal of dyes and heavy metal ions at various initial adsorbate concentrations (Afroze, 2016a; Dawood, 2018)

Adsorbent	Adsorbate	Initial concentration range (mg/L)	Adsorption amount and % removal range	Reference
Fe_3O_2/kaolin composite	Direct Red 23	20–60	46–93%	Magdy et al. (2017)
Weed biochar	Methylene Blue	50–400	27–78 mg/g	Güzel et al. (2017)
Acid-modified sugarcane bagasse	Diazonium Blue	50–250	66–28%	Said et al. (2017)

(Continued)

TABLE 1.3

(Continued)

Adsorbent	Adsorbate	Initial concentration range (mg/L)	Adsorption amount and % removal range	Reference
Bentonite	Al (III)	10–80	1–8.5 mg/g	Chai et al. (2017)
Banana peel biochar	Pb (II)	5–1000	7–270 mg/g	Zhou et al. (2017)
Mango leaf	Methylene Blue	100–250	79–172 mg/g	Uddin et al. (2017)
Chitosan-based activated carbon	Methylene Blue	25–400	27–129 mg/g	Marrakchi et al. (2017)
Chickpea shell	Acid Blue 25	25–100	2.1–7.7 mg/g	Krishna et al. (2017)
Black-eyed biomass	Pb (II)	10–80	95–65%	Guyo and Moyo (2017)
Tea leaf activated carbon	Orange G	200–800	58–99 mg/g	Goswami and Phukan (2017)
Bamboo leaf	Methylene Blue	50–250	23–52 mg/g; 90–42%	Zhu et al. (2016)
Cucumis peel	Acid Blue 113	20–100	12.5–48 mg/g	Lee et al. (2016b)
Hickory wood biochar	Pb (II), Cu (II)	2–100	1.5–17.8 mg/g; 1.7–15.6 mg/g	Ding et al. (2016)
Base-modified eucalyptus bark	Zn (II)	20–70	35–60 mg/g; 69–34%	Afroza et al. (2016c)
Eucalyptus bark	Methylene Blue	20–70	43–78 mg/g; 86–39%	Afroza et al. (2016b)
Kaolin/ iron-manganese oxide composite	Crystal Violet	35–50	16–19 mg/g	Khan et al. (2015)
Rapeseed cake activated carbon	Ni (II), Pb (II)	50–500	40–115 mg/g; 45–122 mg/g	Uçar et al. (2015)
Peanut husk	Indosol Orange	10–200	10–79.5 mg/g	Sadaf and Bhatti (2014)
Pinecone activated carbon	Congo Red	20–60	77–216 mg/g; 77–72%	Dawood et al. (2014b)
Zeolite, Kaolin	Methylene Blue	4–37	3.4–33.5 mg/g; 3.3–30 mg/g	Rida et al. (2013a)
Almond shell char	Ni (II), Co (II)	50–150; 100–200	4.6–5.5 mg/g; 8.6–9.8 mg/g	Kilic et al. (2013)
Coal fly ash	Ni (II), Cu (II)	5–100	100–98%; 96–97.6%	Sočo and Kalembkiewicz (2013)
Raw pinecone	Congo Red	20–50	8.4–24 mg/g	Dawood and Sen (2012)
Kenaf fibre char	Methylene Blue	50–200	10.6–17.8 mg/g	Mahmoud et al. (2012)
Kaolin	Zn (II)	30–50	29–41 mg/g	Aries and Sen (2009)

Generally, the percentage removal of aqueous pollutants increases with contact time before reaching a steady saturation level, often involving a two-stage or multi-stage adsorption process (Dawood & Sen, 2012; Dawood et al., 2018; Sen, 2023). Understanding adsorption kinetics is crucial for determining reaction speed, equilibrium time, and kinetic parameters necessary for adsorber design (Dawood & Sen, 2014b; Sen, 2023).

1.5.3 Effect of Adsorbent Dose on Solid/Liquid Adsorption Process

Understanding the effect of adsorbent dose is essential for designing and scaling up a continuous adsorption column, as adsorption capacity varies with adsorbent quantity. This effect can be studied by preparing an adsorbent–adsorbate solution with varying amounts of adsorbent at a fixed initial adsorbate concentration and solution pH, shaking the mixture until equilibrium is achieved (Yagub et al., 2014). Generally, the percentage removal of organic dyes and inorganic heavy metals increases with rising adsorbent dosage. For example, Kılıç et al. (2013) found that increasing almond shell biochar dosage from 1 to 10 g/L raised Ni(II) and Co(II) ion removal from 10 to 38% and from 25 to 50%, respectively, while the adsorption capacity, q_e (mg/g), decreased from 10 to 3 mg/g for Ni(II) and from 24 to 7 mg/g for Co(II).

The adsorbent dose determines the material's capacity over a range of solute concentrations, with a fixed number of active sites per unit mass. As overall adsorbent mass increases, the available active sites per unit mass decrease, which can reduce adsorption capacity. Moreover, a higher adsorbent dose reduces the concentration gradient between the bulk phase and adsorbent surface, thus reducing the net mass flux. Nonetheless, higher adsorbent doses increase the percentage removal of metal ions due to more available active sites (Afroze et al., 2016b; Dawood et al., 2014b; Maheshwari et al., 2015; Zwain et al., 2014). Some studies, however, report a decrease in percentage removal with increasing adsorbent dose. Senthil Kumar et al. (2014) observed that MB dye removal on modified mango seed decreased from 99.8 to 79% as the adsorbent dose increased from 0.1 to 1.2 g. Similarly, Uddin et al. (2017) found that while MB dye removal increased from 82 to 99%, the adsorption capacity decreased from 164 to 25 mg/g when the mango leaf adsorbent dose increased from 0.1 to 0.8 g (Figure 1.10). A selected list of studies on the effect of adsorbent dose on the percentage removal of metal ions and dyes is presented in Table 1.4.

Figure 1.11 illustrates the variation in removal efficiency for MB dye and tetracycline hydrochloride (TCH) antibiotics with different doses of JA-MIONs (magnetic iron oxide modified with *Justicia adhatoda* leaf extract), ranging from 2 to 10 mg at a fixed initial adsorbate concentration, constant room temperature (25°C), and pH 7. As shown in Figure 1.11, the percentage removal of both MB dye and TCH antibiotics increased with the adsorbent dose, primarily due to the increased availability of adsorption sites (Ghosh et al., 2023).

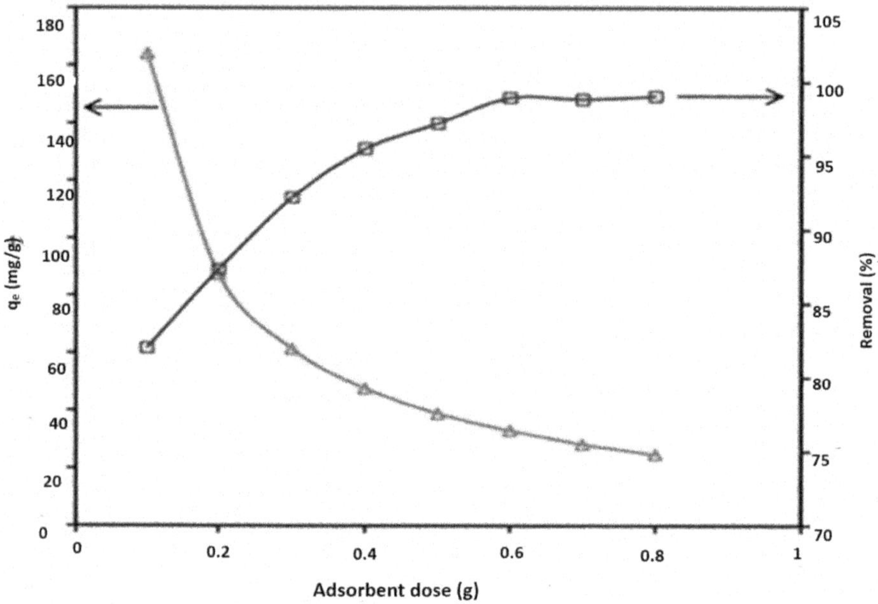

FIGURE 1.10
Effect of adsorbent dose on methylene blue dye adsorption by mango leaf. Initial dye concentration = 100 mg/L, volume = 200 mL (Uddin et al., 2017)

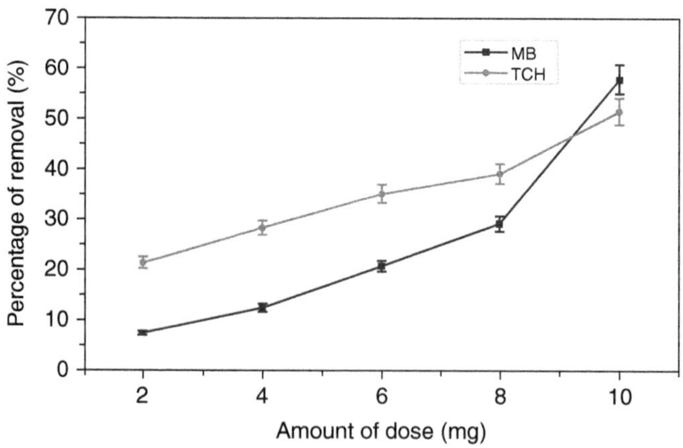

FIGURE 1.11
Effect of adsorbent dose on MB dye and TCH removal using JA-MIONs (Source: Ghosh et al., 2023, with written permission)

TABLE 1.4

Effect of adsorbent dosage on adsorption capacity and percentage removal of dyes and heavy metals

Adsorbent	Adsorbate	Adsorbent dose range	Removal range	Reference
Brassica campestris agricultural waste	Ni^{2+}, Pb^{2+}, Cr^{6+}	0.2–1 g/L	Increases, 200–19 mg/g	Shaike et al. (2018)
Banana peel biochar	Pb (II)	0.2–1 g/L	200–19 mg/g	Zhou et al. (2017)
Mango leaf	Methylene Blue	0.1–0.8 g	60–98%	Uddin et al. (2017)
Fe_3O_2/kaolin composite	Direct Red 23	0.25–0.75 g	79–100%	Magdy et al. (2017)
Base-modified cedar biomass	Ni (II), Pb (II)	0.1–1.5 g	28–8%, 88–23%	Lago et al. (2017)
Chickpea shell	Acid Blue 25	10–300 mg	1–55%	Krishna et al. (2017)
Black-eyed biomass	Pb (II)	0.2–1.6 g/L	77–96%	Guyo and Moyo (2017)
Bamboo leaf	Methylene Blue	0.05–0.3 g	37–93%	Zhu et al. (2016)
Bentonite	Al (III)	4–20 g/L	35–98%	Chai et al. (2017)
Kaolin	Al (III)	4–20 g/L	12–90%	Chai et al. (2017)
Cucumis peel	Acid Blue 113	0–20 g/L	8–78%	Lee et al. (2016b)
Eucalyptus bark (EB)	Zn (II)	10–30 mg	45–30%	Afroza et al. (2016c)
Base-modified eucalyptus bark	Zn (II)	10–30 mg	70–55%	Afroza et al. (2016c)
Eucalyptus bark	Methylene Blue	10–30 mg	73–94%	Afroza et al. (2016b)
Iron-manganese oxide/kaolin composite	Crystal Violet, Basic Fuchsin	1–4 g/L, 2–5 g/L	84–97%, 41–94.5%	Khan et al. (2015)
Tea waste activated carbon	Eosin Yellow	1–3 g/L	88–98%	Borah et al. (2015)
Peanut husk	Indosol Orange	0.05–0.3 g	41–8 mg/g	Sadaf and Bhatti (2014)
Pinecone carbon	Congo Red	10–30 mg	77–81%	Dawood et al. (2014a)
Modified mango seed	Methylene Blue	0.1–1.2 g	69–98%	Senthil Kumar et al. (2014)
Almond shell biochar	Ni (II), Co (II)	1–10 g/L	10–38%, 26–50%	Kılıç et al. (2013)
Zeolite, Kaolin	Methylene Blue	0.1–3 g/L	92–96%, 92–94.5%	Rida et al. (2013a)
Raw pinecone	Congo Red	10–30 mg	13.5–19%	Dawood and Sen (2012)

(Continued)

TABLE 1.4
(Continued)

Adsorbent	Adsorbate	Adsorbent dose range	Removal range	Reference
Acid-modified pinecone	Congo Red	30 mg	85%	Dawood and Sen (2012)
Pine tree leaves	Methylene Blue	10–30 mg	95.5–98%	Yagub et al. (2012)
Kaolin	Zn (II)	10–30 mg	29–10.7 mg/g	Aries and Sen (2009)

1.5.4 Effect of Temperature on Solid/Liquid Adsorption Process

Temperature is a crucial physicochemical process parameter that affects the adsorption or biosorption mechanism, influencing the equilibrium adsorption capacity (Sen, 2023). It also plays a role in determining thermodynamic parameters such as enthalpy, entropy, and free energy, providing insights into whether the adsorption reaction is exothermic or endothermic (Sen, 2023). Different metal ions and adsorbents respond uniquely to temperature changes (Gupta et al., 2010; Monteiro et al., 2010).

Research indicates that metal ion uptake often increases with rising temperature due to factors such as enhanced activity of functional groups, increased affinity of active sites for complex formation with metal ions, and reduced mass transfer resistance (Bayes et al., 2012; Harris, 2010; Sen & Dustin, 2011). Conversely, some studies report a decrease in metal ion uptake with increasing temperature for certain adsorbents (Gupta et al., 2010; Tuzen et al., 2009), while others show minimal or no temperature effect on uptake (Rangsayatorn et al., 2002). In exothermic systems, lower temperatures are generally favourable for enhanced adsorption (Randhawa et al., 2014). Examples of temperature effects on various adsorbent-adsorbate systems are provided in Table 1.5.

1.5.5 Effect of Presence of Salt (Ionic Strength) on Solid/Liquid Adsorption Process

Industrial waste and natural water often contain dissolved salts, with concentrations varying based on wastewater source and quality. The presence of electrolytes in solution significantly influences interfacial potential, competes with adsorbate ions for binding sites, and affects the sorption of adsorbates (such as dyes and heavy metal ions), thereby impacting the adsorption process (Afroza & Sen, 2018). Wastewater, particularly rich in organic and inorganic materials, typically has a higher salt concentration, making the effect of ionic strength important in dye and heavy metal ion adsorption (Afroza & Sen, 2018).

TABLE 1.5

Effect of temperature on different adsorbent-adsorbate solid/liquid adsorption systems

Adsorbent	Adsorbate	Temperature range (K)	Type of process	Reference
Raw pomegranate peel	Cu^{2+}	303–313	Endothermic	Ben-Ali et al. (2017)
Eucalyptus sheathiana bark (raw)	Zn^{2+}	303–333	Exothermic	Afroze et al. (2016c)
Citrus peel waste biomass	Zr(IV)	303–333	Exothermic	Bhatti et al. (2016)
Tomato leaf	Ni^{2+}	303–323	Endothermic	Gutha et al. (2015)
Tomato waste-derived biochar	Cr^{2+}	293–313	Endothermic	Önal et al. (2014)
Dried cactus	Cd^{2+}, Pb^{2+}	298–333	Endothermic	Barka et al. (2013)
Mosambi (*Citrus limetta*) peel	Cr(VI)	303–313	Endothermic	Saha et al. (2013)
Chlorella vulgaris algae	Cd^{2+}	293–323	Exothermic	Aksu (2001)
Maize silk powder	Methylene Blue	298–338	Exothermic	Miraboutalebi et al. (2017)
Cauliflower leaves	Methylene Blue	303–323	Exothermic	Ansari et al. (2016)
Eucalyptus sheathiana bark	Methylene Blue	303–333	Endothermic	Afroze et al. (2016b)
Waste manure ash	Basic Yellow 2	298–318	Endothermic	Öztürk and Malkoc (2015)
Peanut hull	Novacorn Golden Yellow	303–333	Exothermic	Nawaz et al. (2014)
Pomelo (*Citrus grandis*) peel	Reactive Blue 114	283–333	Endothermic	Argun et al. (2014)
Rice husk (nitric acid treated)	Malachite Green	303–333	Endothermic	Ramaraju et al. (2014)
Sawdust	Tartrazine (Acid Yellow 23, E102)	298–318	Endothermic	Banerjee and Chattopadhyaya (2017)
Sugarcane bagasse	Rhodamine B	303–323	Endothermic	Zhang et al. (2013)
Activated bamboo waste	Reactive Black 5	303–323	Endothermic	Ahmad et al. (2013)
Pinecone	Methylene Blue	303–333	Exothermic	Dawood et al. (2017a)

TABLE 1.6

Effect of salt/electrolyte concentration on adsorption of heavy metals and organic dyes

Adsorbent	Adsorbate	Salt (NaCl/ $CaCl_2$/$FeCl_3$) concentration (mg/L)	Percent removal	Reference
Cedar bark	Cu^{2+}	0.1–1 g/L	Decrease	Hamdaoui (2017)
Mango kernel (biocomposite)	Cr(VI)	100	Decrease	Akram et al. (2017)
Eucalyptus sheathiana bark (modified)	Zn^{2+}	100–300	62.83–54.19	Afroze et al. (2016c)
Eucalyptus bark	Cd^{2+}	0–10 g/L (Na_2SO_4)	Decrease	Ghodbane et al. (2008)
Ulmas tree leaves	Pb^{2+}, Cu^{2+}, Cd^{2+}	0–0.1 M	Decrease	Sangi et al. (2008)
Eucalyptus bark	Hg^{2+}	0–20 g/500 mL	Decrease	Ghodbane and Hamdaoui (2008)
Maize silk powder	Methylene Blue	20–180	39.3–80.7	Miraboutalebi et al. (2017)
Eucalyptus sheathiana bark biomass	Methylene Blue	100–300 mg/L	58.78–69.92	Afroze et al. (2016b)
Cauliflower leaves	Methylene Blue	0.0–0.1 M	Decrease	Ansari et al. (2016)
Modified breadnut peel	Malachite Green	0.0–1.0 M (KNO_3)	Decrease	Chieng et al. (2015)
Modified phoenix tree leaves	Crystal Violet	0.0–0.1 M	Decrease	Ren et al. (2013)
Cinnamomum camphora sawdust	Malachite Green	0–0.05	89.21–64.75	Wang et al. (2014)
Modified peanut husk	Light Green	0–0.1	Decrease	Zhao et al. (2014)
Pirina (treated with nitric acid)	Remazon Brilliant Blue	0.05–0.3	96.45–86.15	Dağdelen et al. (2014)
Pleurotus ostreatus	Malachite Green	0.05–0.2 M	87.22–49.01	Chen et al. (2014)
Lotus leaf	Malachite Green	0–0.2 M	81.7–62.5	Han et al. (2014)
Pinecone	Methylene Blue	0–200	Decrease	Sen et al. (2011)
Potato peel	Malachite Green	0–2.0 g	Decrease	Guechi and Hamdaoui (2016)

To assess the impact of ionic strength on adsorption capacity and adsorbate removal percentage, researchers prepared adsorbate-adsorbent solutions with varying initial salt concentrations while keeping other parameters constant (solution temperature, contact time, solution pH, and stirring speed) until equilibrium was reached (Afroze et al., 2016c; Stavropoulos et al., 2015). Numerous studies have investigated how co-ions affect adsorbent efficiency. In laboratory experiments, common salts used include monovalent (NaCl), divalent ($CaCl_2$), and trivalent ($FeCl_3$), as these are frequently found in industrial effluents.

Results indicate that higher salt concentrations generally have an inverse effect on adsorption efficiency. External electrolytes, such as Na^+, K^+, and Ca^{2+}, can limit the binding efficiency between adsorbent and adsorbate, as ions may compete for adsorption sites, thereby reducing adsorption capacity. This phenomenon is partly due to changes in the equilibrium constant between the bulk liquid and the interface as salt concentration increases, affecting the overall adsorption process. Additionally, factors like molecular weight, atomic size, radii, and charge density of ions influence adsorption efficiency. Divalent ions have a stronger impact on ionic strength and positive charge than monovalent ions, making their influence on the adsorption process more pronounced (Afroze et al., 2016c; Dawood, 2018).

Divalent and trivalent ions reduce adsorption rates more effectively than monovalent ions. Transition metals, such as Fe, exhibit greater affinity for adsorbent surfaces than alkaline earth metals like Ca (Afroze et al., 2016b; Biswas et al., 2019b). Examples of how various salt/electrolyte concentrations affect the adsorption of heavy metals and organic dyes are presented in Table 1.6 (Afroze, 2016a).

1.6 Air Pollutant Removal by Gas/Solid Adsorption

Industrialization, urbanization, fossil fuel combustion, and agricultural activities generate a range of air pollutants, including carbon dioxide (CO_2), sulphur dioxide (SO_2), hydrogen sulphide (H_2S), volatile organic compounds (VOCs), and ammonia (NH_3). These pollutants pose significant risks to the environment and human health (Sen, 2017; Zhao et al., 2022). In addition to conventional methods such as membrane separation for H_2S and CO_2, absorption for CO_2 and VOCs, desulphurization and wet washing for SO_2, and advanced oxidation technologies (AOTs), the adsorption method has gained prominence for air pollutant removal (Sen, 2017; Zhao et al., 2022). Various adsorbents, including activated carbon, carbon nanotubes, graphene derivatives, nanomaterials, zeolites, metal-organic frameworks, polymeric materials, composites, and biochar, are commonly used for gas/solid adsorption of air pollutants.

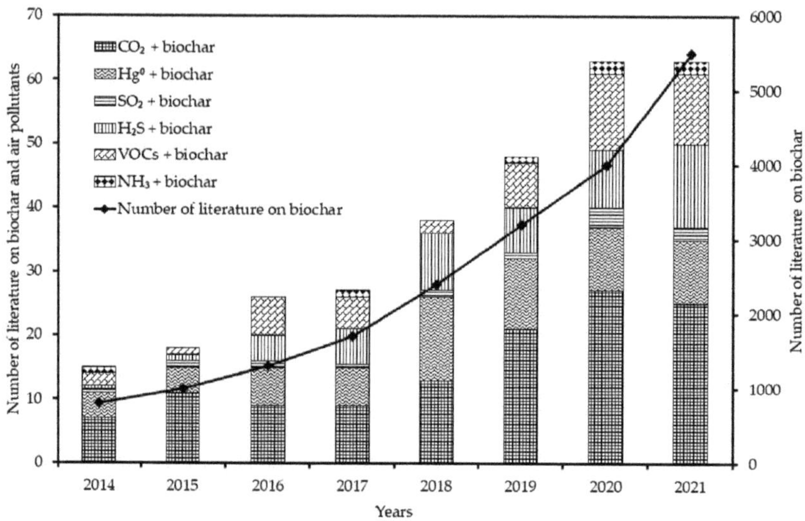

FIGURE 1.12
Number of studies on biochar and its removal of six air pollutants from 2014 to 2021 (Data from Web of Science, Source: Zhao et al., 2022, with permission)

Biochar has attracted significant interest as a cost-effective adsorbent for industrial air pollutants (Zhao et al., 2022). Figure 1.12 illustrates the rising number of studies from 2014 to 2021 on biochar's effectiveness in removing six major air pollutants. Zhao et al. (2022) provided a comprehensive review of biochar feedstocks, synthesis, and activation methods and its application in adsorptive removal of six key air pollutants—SO_2, H_2S, CO_2, VOCs, Hg^0, and NH_2—under varying conditions (Figure 1.13). The possible mechanisms of gaseous pollutant adsorption on biochar are also highlighted in Figure 1.13.

Fly ash (FA), a cost-effective industrial solid waste-based adsorbent, has proven effective in adsorbing air pollutants due to its high porosity, surface area, and oxide content, as well as the presence of unburned carbon and CaO in its particles (Ge et al., 2018; Sen, 2017). Fly ash can adsorb pollutants like VOCs, NO_x, and SO_2 (Ge et al., 2018).

Activated carbon from biodegradable waste, with a surface area of 1426 m^2/g, has shown high sorption capacities for nitrogen dioxide (NO_2) and hydrogen sulphide (H_2S) at 83 and 215 mg/g, respectively (Nowicki et al., 2016). Additionally, waste ion exchange resin-based microporous activated carbon achieved a CO_2 adsorption capacity of 81.24 mg CO_2/g at 30°C under optimal activation conditions (Wei et al., 2016). Marigold-based activated carbon has also been effective in adsorbing NO_2, reaching a maximum adsorption capacity of 102.1 mg/g under specific activation conditions (Bazan et al., 2016).

Liu et al. (2021) reviewed sulphate radical ($SO_4{\bullet}^-$)-based AOTs for removing gaseous pollutants, highlighting them as a promising technology. Biofiltration

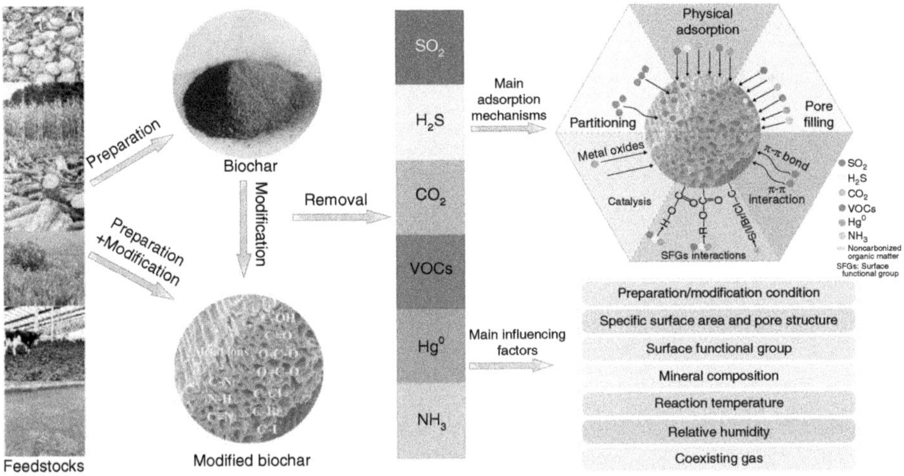

FIGURE 1.13
Graphical representation of biochar types, synthesis methods, and applications in air pollutant removal by gas/solid adsorption, highlighting adsorption mechanisms.

Source: Zhao et al., 2022, with permission

has also gained traction as a low-cost alternative for removing odours, VOCs, and other air pollutants due to its simplicity, high reduction efficiency, and minimal energy requirements (Sheoran et al., 2022).

1.7 Regeneration and Reusability of Spent Adsorbents

Desorption technology is essential for recovering adsorbed molecules from spent adsorbents and enabling adsorbent reuse, which enhances the efficiency and economic feasibility of adsorption technology (Bayuo et al., 2023). The advantages of spent adsorbent regeneration and the challenges of disposal are illustrated in Figure 1.14 (Vakili et al., 2019).

The desorption process typically involves mixing an appropriate solvent with the metal or dye-saturated adsorbent and shaking the mixture until the solute detaches from the adsorbent into the solvent, after which filtration separates the adsorbent (Afroza & Sen, 2018). Gautam et al. (2014) reviewed various eluents, including EDTA, NaOH, Na_2SO_4, HCl, H_2SO_4, and HNO_3, and their concentrations for recovering pollutants from loaded adsorbents. Factors such as desorption performance, safety, economic viability, and environmental friendliness guide the selection of these eluents.

Adsorbent regeneration techniques fall into thermal, biological, ultrasonic, and chemical categories. The regenerated adsorbents can be reused for

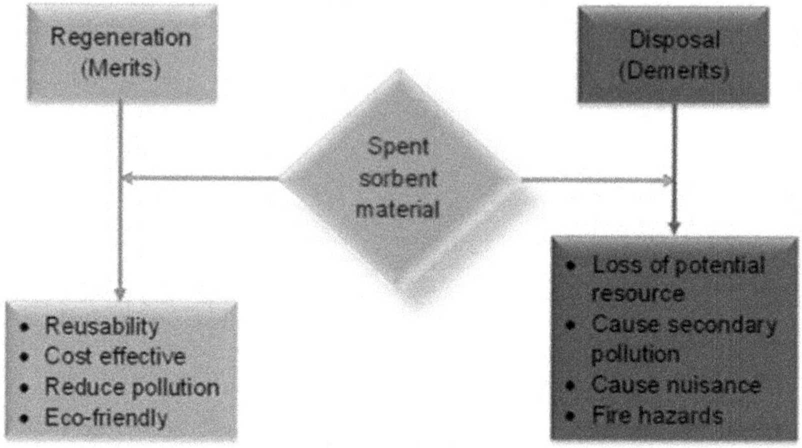

FIGURE 1.14
Advantages of regenerating spent adsorbents and issues related to storage/disposal

Source: Vakili et al., 2019, with permission.

further adsorption processes or recovery of valuable chemicals. They are also applicable in developing biosensors, enhancing the versatility and sustainability of the adsorption process.

Bibliography

Abdolali, A., H. H. Ngo, W. Guo, S. Lu, S. S. Chen, N. C. Nguyen, X. Zhang, J. Wang, Y. Wu (2016). A breakthrough biosorbent in removing heavy metals: Equilibrium, kinetic, thermodynamic and mechanism analyses in a lab-scale study. *Science Total Environment*, 542, 603–611. https://doi.org/10.1016/j.scitotenv.2015.10.095

Afroze, S. (2016a). Ph.D thesis entitled aqueous phase adsorption of organic/inorganic contaminants by eucalyptus bark (Eucalyptus sheathiana) biomass. Curtin University, Perth, AUSTRALIA. URI: https://hdl.handle.net/20.500.11937/1561

Afroza, S., T. K. Sen, M. Ang (2016b). Adsorption Performance of Continuous Fixed Bed Column for the removal of Methylene Blue (MB) Dye using Eucalyptus Sheathiana Bark biomass. *Research Chemical Intermediate*, 42, 2343–2364. https://doi.org 10.1007/s11164-015-2153-8

Afroza, S., T. K. Sen, M. Ang (2016c). Adsorption removal of Zinc (II) from Aqueous Phase by Raw and Base Modified *Eucalyptus Sheathiana* Bark: Kinetics, mechanism and equilibrium study. *Process Safety and Environmental Protection*, 102, 336–357. https://doi.org10.1016/j.psep.2016.04.009

Afroze, S., T. K. Sen (2018). Removal of aqueous phase heavy metal ions and dyes using agricultural solid wastes adsorbents: A review. *Water, Air & Soil Pollution*, 229(225), 1–50. https://doi.org/10.1007/s11270-018-3869-z

Ahmad, A., Z. A. Ghazi, M. Saeed, M. Ilyas, R. Ahmad, A. Muqsit Khattak, A. Iqbal (2017). Comparative study of the removal of Cr(vi) from synthetic solution using natural biosorbents. *New Journal Chemistry*, 41, 10799–10807. https://doi.org/10.1039/C7NJ02026K

Ahmad, T., M. A. Danish (2022). Review of avocado waste-derived adsorbents: Characterizations, adsorption characteristics, and surface mechanism. *Chemosphere*, 296, 134036. https://doi.org/10.1016/j.chemosphere.2022.134036

Ahmad, A. A., A. Idris, B. H. Hameed (2013). Organic dye adsorption on activated carbon derived from solid waste. Desalin. *Water Treatment*, 51, 2554–2563. https://doi.org/10.1080/19443994.2012.749019

Akram, M., H. N. Bhatti, M. Iqbal, S. Noreen, S.Sadaf (2017). Biocomposite efficiency *for Cr(VI)* adsorption: Kinetic, equilibrium and thermodynamics studies. *Journal of Environmental Chemical Engineering*, 5, 400–411. https://doi.org/10.1016/j.jece.2016.12.002

Aksu, Z. (2001). Equilibrium and kinetic modelling of cadmium(II) biosorption by C. vulgaris in a batch system: Effect of temperature. *Separtion and Purification Technology*, 21, 285–294. https://doi.org/10.1016/S1383-5866(00)00212-4

Angove, M. J., B. B. Johnson J. D. Wells (1998). The influence of temperature on the adsorption of cadmium (II) and cobalt(II) on kaolinite. *Journal of Colloid and Interface Science*, 204, 93–103.

Ansari, S. A., F. Khan, A. Ahmad (2016). Cauliflower leave, an agricultural waste biomass adsorbent, and its application for the removal of MB dye from aqueous solution: Equilibrium, kinetics, and thermodynamic studies. *International Journal Analytical Chemistry*, 2016, 1–10. https://doi.org/10.1155/2016/8252354

Apark, R., K. Guclu, M. H. Turgut (1998). Modeling of copper (II), cadmium (II), and lead (II) adsorption on red mud. *Jorunal of Colloid and Interface Science*, 203, 122–130.

Argun, M. E., D. Güclü, M. Karatas (2014). Adsorption of Reactive Blue 114 dye by using a new adsorbent: Pomelo peel. *Journal of Industrial and Engineering Chemistry*, 20, 1079–1084. https://doi.org/10.1016/j.jiec.2013.06.045

Arias, F., T. K. Sen (2009). Removal of zinc metal ion (Zn^{2+}) from its aqueous solution by kaolin clay minerals: A kinetic and equilibrium study. *Colloids and Surfaces A*, 348, 100–108.

Asadi, E., P. Haroonian, M. Ghaedi, A. Asfaram (2024). Removal of tartrazine and indigo carmine with layered double hydroxide-modified graphite nanocomposite: Isotherm, kinetics and response surface methodology (RSM) optimization. *Journal of Molecular Liquids*, 402, 124769. https://doi.org/10.1016/j.molliq.2024.124769

Banerjee, S., M. C. Chattopadhyaya (2017). Adsorption characteristics for the removal of a toxic dye, tartrazine from aqueous solutions by a low cost agricultural by-product. *Arabian Journal of Chemistry*, 10, S1629–S1638. https://doi.org/10.1016/j.arabjc.2013.06.005

Barakat, M. A., E. Schmidt (2010). Polymer-enhanced ultrafiltration process for heavy metals removal from industrial wastewater. *Desalination*, 256, 90–93. https://doi.org/10.1016/j.desal.2010.02.008

Barka, N., M. Abdennouri, M. El Makhfouk, S. Qourzal (2013). Biosorption characteristics of cadmium and lead onto eco-friendly dried cactus (Opuntia ficus indica) cladodes. *Journal Environmental Chemical Engineering*, 1, 144–149. https://doi.org/10.1016/j.jece.2013.04.0

Bayes, G. S., S. S. Raut, V. R. Patil, R. S. Lokhande (2012). Formation of diazepam-lanthanides(III) complexes in the 50-50 volume % ethanol-water solvent system and study of the effect of temperature on the complex formation constants. *Journal Solution Chemistry*, 41, 241–248. https://doi.org/10.1007/s10953-012-9798-3

Bayuo, J., M. J. Rwiza, K. M. Mtei (2023). Non-competitive and competitive detoxification of As(III) ions from single and binary biosorption systems and biosorbent regeneration. *Biomass Conversion and Biorefinery*, 14, 1–28. https://doi.org/10.1007/s13399-022-03734-0

Bazan, A., P. Nowicki, R. Pietrzak (2016). Removal of NO2 by carbonaceous adsorbents obtained from residue after supercritical extraction of marigold. *Adsorption*, 22, 465–471. https://doi.org/10.1007/s10450-015-9709-1

Ben-Ali, S., I. Jaouali, S. Souissi-Najar, A. Ouederni (2017). Characterization and adsorption capacity of raw pomegranate peel biosorbent for copper removal. *Journal of Cleaner Production*, 142, 3809–3821. https://doi.org/10.1016/j.jclepro.2016.10.081

Benjamin, M. M., J. Leckie (1981). Multiplesite adsorption of Cd, Cu, Zn and Pb on amorphous iron oxyhydroxide. *Journal of Colloid Interface Science*, 79, 209–219.

Bhatti, H. N., Q. Zaman, A. Kausar, S. Noreen, M. Iqbal (2016). Efficient remediation of Zr(IV) using citrus peel waste biomass: Kinetic, equilibrium and thermodynamic studies. *Ecological Engineering*, 95, 216–228. https://doi.org/10.1016/j.ecoleng.2016.06.08

Biswas, S., H. Siddiqi, B. C. Meikap, T. K. Sen, M. Khiadani (2020a). Preparation and characterization of raw and inorganic acid-activated pine cone biochar and its application in the removal of aqueous-phase Pb^{2+} metal ions by adsorption. *Water, Air, and Soil Pollution*, 231(1), 3.

Biswas, S., S. S. Mohapatra, U. Kumari, B. C. Meikap,T. K. Sen (2020b). Batch and continuous closed circuit semi-fluidized bed operation: Removal of MB dye using sugarcane bagasse biochar and alginate composite adsorbents. *Journal of Environmental Chemical Engineering*, 8(1), 103637. https://doi.org/10.1016/j.jece.2019.103637

Biswas, S., R. K. Diwakar, Ipshita D. Behera, B. C. Meikap, T. K. Sen, M. Khiadani (2020c). Aqueous phase phenol removal from synthetic and real steel plant effluents through a batch and Semifluidized bed clumn operation: Experimental and model analysis. *Journal Environmental Chemical Engineering*, 8, 104441. https://doi.org/10.1016/j.jece.2020.104441

Biswas, S., M. Bal, S. K. Behera, T. K. Sen, B. C. Meikap (2019a). Process optimization study of Zn2+ adsorption on biochar-alginate composite adsorbent by Response Surface Methodology (RSM). *Water*, 11(2), 325. https://doi.org/10.3390/w11020325

Biswas, S., T. K. Sen, A. M. Yeneneh, B. C. Meikap (2019b). Synthesis and characterization of a novel Ca-alginate-biochar composite as efficient zinc (Zn^{2+}) adsorbent: Thermodynamics, process design, mass transfer and isotherm modelling. *Separation Science and Technology*, 54(7), 1106–1124.

Biswas, S., B. C. Meikap, T. K. Sen (2019c). Adsorptive removal of aqueous phase copper (Cu2+) and nickel (Ni2+) metal ions by synthesize Biochar-biopolymeric hybrid adsorbents and process optimization by response surface methodology (RSM). *Water, Air & Soil Pollution*, 230, 197. https://doi.org/10.1007/s11270-019-4258-y

Borah, L., M. Goswami, P. Phukan (2015). Adsorption of methylene blue and eosin yellow using porous carbon prepared from tea waste: Adsorption equilibrium, kinetics and thermodynamics study. *Journal of Environmental Chemical Engineering*, 3(2), 1018–1028.

Boruah, P., A. Sarma, K. G. Bhattacharyya (2015). Removal of Ni (II) ions from aqueous solution by using low cost biosorbent prepared from jackfruit (Artocarpus heterophyllus) leaf powder. *Indian Journal of Chemical Technology*, 22(6), 322–327.

Borah, L., M. Goswami, P. Phukan (2015). Adsorption of methylene blue and eosin yellow using porous carbon prepared from tea waste: Adsorption equilibrium, kinetics and thermodynamics study. *Journal of Environmental Chemical Engineering*, 3(2), 1018–1028.

Burakova, A. E., E. V. Galunina, I. V. Burakovaa, A. E. Kucherovaa, S. Agarwalb, A. G. Tkacheva, V. K. Guptab (2018). Adsorption of heavy metals on conventional and nanostructured materials for wastewater treatment purposes: A review. *Ecotoxicology and Environmental Safety*, 148(2018), 702–707. https://doi.org/10.1016/j.ecoenv.2017.11.034

Chai, W., Y. Huang, G. Han, J. Liu, S. Yang, Y. Cao (2017). An enhanced study on adsorption of Al(iii) onto Bentonite and Kaolin: Kinetics, Isotherms, and Mechanisms. *Mineral Processing and Extractive Metallurgy Review*, 38(2), 106–115.

Chen, H., J. Zhao, G. Dai, J. Wu, H. Yan (2010). Adsorption characteristics of Pb(II) from aqueous solution onto a natural biosorbent, fallen Cinnamomum camphora leaves. *Desalination*, 262(1–3), 174–182. https://doi.org/10.1016/j.desal.2010.06.006

Chen, L., H.-X. Zhang (2022). Modified biochar adsorbents (MBAs) for heavy metal ions adsorption: A critical review. *Journal of Environmental Chemical Engineering*, 10(2), 107393.

Chen, Z., H. Deng, C. Chen, Y. Yang, H. Xu (2014). Biosorption of malachite green from aqueous solutions by Pleurotus ostreatus using Taguchi method. *Journal Environmental Health Science and Engineering*, 12, 63. https://doi.org/10.1186/2052-336X-12-63

Chieng, H. I., L. B. L. Lim, N. Priyantha (2015). Enhancing adsorption capacity of toxic malachite green dye through chemically modified breadnut peel: Equilibrium, thermodynamics, kinetics and regeneration studies. *Environmental Technology*, 36, 86–97. https://doi.org/10.1080/09593330.2014.938124

Cornell, R. M., U. Schwertmann (1998). *The Iron Oxide*, 1st edition. VCH Publishers.

Crini, G. (2006). Non-conventional low-cost adsorbents for dye removal: A review. *Bioresource Technology*, 97, 1061–1085. https://doi.org/10.1016/j.biortech.2005.05.001

Dağdelen, S., B. Acemioğlu, E. Baran, O. Koçer (2014). Removal of Remazol Brilliant Blue R From aqueous solution by pirina pretreated with nitric acid and commercial activated carbon. *Water, Air, Soil Pollution*, 225, 1899. https://doi.org/10.1007/s11270-014-1899-8

Dawood, S., T. K. Sen (2012). Removal of anionic dye Congo red from aqueous solution by raw pine and acid treated pinecone powder as adsorbent: Equilibrium, thermodynamic, kinetics, mechanism and process design. *Water Research*, 46, 1933–1946.

Dawood, S. (2018). Synthesis and characterization of biomass and clay minerals based adsorbents for the removal of cationic dye and metal ion from wastewater by adsorption. *Ph.D. thesis*. 2018, Curtin University Liberary, Bentley, WA 6845, AUSTRALI. URI. https://hdl.handle.net/20.500.11937/69346.

Dawood, S., T. K. Sen, C. Pan (2018). Performance and dynamic modelling of biochar and kaolin packed bed adsorption column for aqueous phase methylene blue (MB) dye removal. *Environmental Technology,* 40, 1–11. https://doi.org/10.1080/09593330.2018.1491065

Dawood, S., T. Gupta, T. K. Sen (2017a). Adsorptive removal of methylene blue (MB) dye at kaolin clay-water interface: Kinetics, isotherm modelling and process design, Chapter 7, in: Book entitled *"Clay Minerals: Properties, Occurrence and Uses,* edited by T. K. Sen, Nova Publishers Inc ISBN 978-1-53610-823-1, pp. 209–236.

Dawood, S., T. K. Sen, C. Phan (2017b). Synthesis and Characterisation of Pine Cone bio-char in the removal of organic and inorganic pollutants from aqueous solution by adsorption: Kinetic, equilibrium, mechanism and thermodynamic. *Bioresource and Biotechnology,* 246, 76–78. https://doi.org/10.1016/j.biortech.2017.07.019

Dawood, S., T. K. Sen, C. Phan (2014a). Preparation and characterization of novel activated carbon from waste biomass pine cone and its application in Congo red dye adsorption from aqueous solution: Equilibrium, thermodynamic, kinetics, mechanism and process design. *Water, Air & Soil Pollution,* 225, 1818. https://doi.org/10.1007/s11270-013-1818-4

Dawood, S., T. K. Sen (2014b). Review on dye removal from its aqueous solution into alternative cost-effective non-conventional adsorbents. *Journal Chemical and Process Engineering,* 1, 1–7.

Deliyanni, E. A., G. Z. Kyzas, K. S. Triantafyllidis, K. A. Matis (2015). Activated carbons for the removal of heavy metal ions: A systematic review of recent literature focused on lead and arsenic ions. *Open Chemistry,* 13, 699–708. From the *journal Open Chemistry*. https://doi.org/10.1515/chem-2015-0087

Deng, Z., K. Y. Fung, K. M. Ng, C. Wei (2016). Design of anaerobic fluidized bed bioreactor – Dyeing effluents. *Chemical Engineering Science,* 139(Supplement C), 273–284.

Du, Q., Z. Sun, W. Forslins, H. Tang (1997). Adsorption of copper at aqueous illite surfaces. *Journal Colloid and Interface Science,* 187, 232–242.

Dutta, B. K. (2007). Principle of mass transfer and separation process. PHI learning, India, 1st edition, ISBN-10: 8120329902, ISBN-13: 978–8120329904.

Dzombak, D. A., F. M. M. Morel (1990). *Surface Complexing Modelling. Hydrous Ferric Oxide*. John Wiley 393–393.

Dzombak, D., F. M. M. Morel (1988). Sorption of cadmium on hydrous ferric oxide at high sorbate/sorbent ratios: Equilibrium, kinetics, and modelling. *Journal of Colloid Interface Science,* 112, 589–597.

El-Baz, A. A., I. A. Hendyb, A. M. Dohdohc, M. I. Soru (2020). Adsorption technique for pollutants removal; current new trends and future challenges – A Review. *The Egyptian International Journal of Engineering Sciences and Technology,* 32, 1–24. https://eijest.journals.ekb.eg. https://eijest.journals.ekb.eg/article_144052_691e2a994301072a8d92e43144eef9d7.pdf

Fu, F., Q. Wang (2011). Removal of heavy metal ions from wastewaters: A review. *Journal of Environmental Management,* 92(3), 407–418. https://doi.org/10.1016/j.jenvman.2010.11.011

Foo, K.Y., B. H. Hameed (2011a). Microwave assisted preparation of activated carbon from pomelo skin for the removal of anionic and cationic dyes. *Chemical Engineering Journal,* 173(2), 385–390.

Gautam, R. K., A. Mudhoo, G. Lofrano, M. C. Chattopadhyaya (2014). Biomass-derived biosorbents for metal ions sequestration: Adsorbent modification and activation methods and adsorbent regeneration. *Journal of Environmental Chemical Engineering*, 2(1), 239–259.

Ge, J. C., S. K. Yoon, N. J. Choi (2018). Application of fly ash as an adsorbent for removal of air and water pollutants. *Applied Science*,8, 1116. https://doi.org/10.3390/app8071116

Geankoplis, C. J. (2003). *Transport Processes and Separation Process Principles (Includes Unit Operation)*. 4th edition. Prentice Hall. ISBN: 013101367X

Ghodbane, I., L. Nouri, O. Hamdaoui, M. Chiha (2008). Kinetic and equilibrium study for the sorption of cadmium(II) ions from aqueous phase by eucalyptus bark. *Journal Hazardous Material*, 152, 148–158. https://doi.org/10.1016/j.jhazmat.2007.06.079

Ghodbane, I., O. Hamdaoui (2008). Removal of mercury(II) from aqueous media using eucalyptus bark: Kinetic and equilibrium studies. *Journal Hazardous Material*, 160, 301–309. https://doi.org/10.1016/j.jhazmat.2008.02.116

Ghosh, N., S. Sen, G. Biswas, A. Saxena, P. K. Haldar (2023). Adsorption and desorption study of reusable magnetic iron oxide nanoparticles modifed with Justicia adhatoda leaf extract for the removal of textile dye and Antibiotic. *Water Air, & Soil Pollution*, (2023)234, 202. https://doi.org/10.1007/s11270-023-06217-8

Gisi, S. D., G. Lofrano, M. Grassi, M. Notarnicola (2016). Characteristics and adsorption capacities of low-cost sorbents for wastewater treatment: A review. *Sustainable Materials and Technologies*, 9, 10–40.

Goswami, M., P. Phukan (2017). Enhanced adsorption of cationic dyes using sulfonic acid modified activated carbon. *Journal of Environmental Chemical Engineering*, 5(4), 3508–351.

Guechi, E.-K., O. Hamdaoui (2016). Sorption of malachite green from aqueous solution by potato peel: Kinetics and equilibrium modeling using non-linear analysis method. *Arabian Journal of Chemistry*, 9, S416–S424. https://doi.org/10.1016/j.arabjc.2011.05.011

Gupta, V. K., A. Rastogi, A. Nayak (2010). Biosorption of nickel onto treated alga (Oedogonium hatei): Application of isotherm and kinetic models. *Journal Colloid Interface Science*, 342, 533–539. https://doi.org/10.1016/j.jcis.2009.10.074

Gutha, Y., V. S. Munagapati, M. Naushad, K. Abburi (2015). Removal of Ni(II) from aqueous solution by Lycopersicum esculentum (Tomato) leaf powder as a low-cost biosorbent. *Desalination and Water Treatment*, 54, 200–208. https://doi.org/10.1080/19443994.2014.880160

Guyo, U., M. Moyo (2017). Cowpea pod (Vigna unguiculata) biomass as a low-cost biosorbent for removal of Pb(II) ions from aqueous solution. *Environmental Monitoring and Assessment*, 189(2), 47.

Güzel, F., H. Sayğılı, G. Akkaya Sayğılı, F. Koyuncu, C. Yılmaz (2017).Optimal oxidation with nitric acid of biochar derived from pyrolysis of weeds and its application in removal of hazardous dye methylene blue from aqueous solution. *Journal of Cleaner Production*, 144(Supplement C), 260–265.

Hamdaoui, O. (2017). Adsorption of Cu(II) from aqueous phase by Cedar bark. *Journal Dispersion Science and Technology*, 38, 1087–1091. https://doi.org/10.1080/01932691.2016.1225261

Han, X., J. Yuan, X. Ma (2014). Adsorption of malachite green from aqueous solutions onto lotus leaf: Equilibrium, kinetic, and thermodynamic studies. *Desalination and Water Treatment*, 52, 5563–5574. https://doi.org/10.1080/19443994.2013.813102

Harris, D. C. (2010). Summary for Policymakers, in: Intergovernmental Panel on Climate Change (Ed.), Climate Change 2013- The Physical Science Basis. Cambridge University Press, Cambridge, pp. 1–30. https://doi.org/10.1017/CBO9781107415324.004

Kamari, S., A. Shahbazi (2020). Biocompatible Fe3O4@SiO2-NH2 nanocomposite as a green nanofiller embedded in PES–nanofiltration membrane matrix for salts, heavy metal ion and dye removal: Long–term operation and reusability tests. *Chemosphere,* 243, 125282. https://doi.org/10.1016/j.chemosphere.2019.125282

Karamveer, S. S., S. Siwal, D. Kapoor, N. Singh, A. K. Saini, W. F. Alsanie, V. K. Thakur (2022). Air pollutants removal using biofiltration technique: A challenge at the frontiers of sustainable environment. *ACS Engineering Au* 2022 2(5), 378–396. https://doi.org/10.1021/acsengineeringau.2c00020

Khalid, K., W. S. Ngah, M. A. Hanafiah, N. S. Malek, N.S., S. N. Khazaai (2015). Acid Blue 25 adsorption onto Phosphoric acid treated Rubber Leaf Powder. *American Journal of Environmental Engineering,* 5(3A), 19–25.

Khan, T. A., E. A. Khan, Shahjahan (2015). Removal of basic dyes from aqueous solution by adsorption onto binary iron-manganese oxide coated kaolinite: Non-linear isotherm and kinetics modeling. *Applied Clay Science,* 107(Supplement C), 70–77.

Kılıç, M., C. Kırbıyık, O. Çepelioğullar, A. E. Pütün (2013). Adsorption of heavy metal ions from aqueous solutions by biochar, a by-product of pyrolysis. *Applied Surface Science,* 283, 856–862.

Kołodyńska, D., J. Krukowska, P. Thomas (2017a). Comparison of sorption and desorption studies of heavy metal ions from biochar and commercial active carbon. *Chemical Engineering Journal,* 307(Supplement C), 353–363.

Kołodyńska, D., J. Krukowska-Bąk, J. Kazmierczak-Razna, R. Pietrzak (2017b). Uptake of heavy metal ions from aqueous solutions by sorbents obtained from the spent ion exchange resins. *Microporous and Mesoporous Materials,* 244(Supplement C), 127–136.

Krishna, L. S., A. Yuzir, G., Yuvaraja, V. Ashokkumar (2017). Removal of Acid Blue25 from aqueous solutions using Bengal gram fruit shell (BGFS) biomass. *International Journal of Phytoremediation,* 19(5), 431–438.

Krishnan, K. A., K. Sreejalekshmi, V. Vimexen, V. V. Dev (2016). Evaluation of adsorption properties of sulphurised activated carbon for the effective and economically viable removal of Zn (II) from aqueous solutions. *Ecotoxicology and Environmental Safety,* 124, 418–425.

Lee, K. M., C. W. Lai, K. S. Ngai, J. C. Juan (2016a). Recent developments of zinc oxide based photocatalyst in water treatment technology: A review. *Water Research,* 88, 428–448.

Lee, L. Y., S. Gan, M. S. Yin Tan, S. S. Lim, X. J. Lee, Y. F. Lam (2016b). Effective removal of Acid Blue 113 dye using overripe Cucumis sativus peel as an eco-friendly biosorbent from agricultural residue. *Journal of Cleaner Production,* 113 (Supplement C), 194–203.

Liang, F. B., Y. L. Song, C. P. Huang, J. Zhang, B. H. Chen (2013). Adsorption of hexavalent chromium on a lignin-based resin: Equilibrium, thermodynamics, and kinetics. *Journal of Environmental Chemical Engineering,* 1(4), 1301–1308. https://doi.org/10.1016/j.jece.2013.09.025

Liu, Q, Y. Zhou, J. Lu, D. Zhou (2020). Novel cyclodextrin-based adsorbents for removing pollutants from wastewater: A critical review. *Chemosphere,* 241, 125043.

Liu, Y, L. Liu, Y. Wang (2021). A critical review on removal of gaseous pollutants using sulfate radical-based advanced oxidation techniques. *Environmental Science and Technnology*, 55(14), 9691–9710. https://doi.org/10.1021/acs.est.1c0153

Lothenbach, B., G. Furrer, R. Schulin (1997). Immobilization of heavy metals by polynuclear aluminum and montmorillonite compounds. *Environmental Science Technology,* 31, 1452–1462.

Magdy, A., Y. O. Fouad, M. H. Abdel-Aziz, A. H. Konsowa (2017). Synthesis and characterization of Fe3O4/kaolin magnetic nanocomposite and its application in wastewater treatment. *Journal of Industrial and Engineering Chemistry,* 56(Supplement C), 299–311.

Mallampati, R., L. Xuanjun, A. Adin, S. Valiyaveettil (2015). Fruit peels as efficient renewable adsorbents for removal of dissolved heavy metals and dyes from water. *ACS Sustainable Chemistry and Engineering*, 3, 1117–1124.

Marrakchi, F., M. J. Ahmed, W. A. Khanday, M. Asif, B. H. Hameed (2017). Mesoporous-activated carbon prepared from chitosan flakes via single-step sodium hydroxide activation for the adsorption of methylene blue. *International Journal of Biological Macromolecules,* 98, 233–239.

Mckenzie, R. M. (1980). The adsorption of lead and other heavy metals on oxides of manganise and iron. *Australian Journal Soil Research,* 18, 61–73.

Miraboutalebi, S. M., S. K. Nikouzad, M. Peydayesh, N. Allahgholi, L. Vafajoo, G. McKay (2017). Methylene blue adsorption via maize silk powder: Kinetic, equilibrium, thermodynamic studies and residual error analysis. *Process Safety and Environmental Protection,* 106, 191–202. https://doi.org/10.1016/j.psep.2017.01.010

Mohammed, M., T. K. Sen, S. Maitra, B. K. Dutta (2011). Removal of Zn^{2+} from aqueous solution using Castor seed hull. *Water Air and Soil Pollution,* 215(1–4), 609–620. https://doi.org/10.1007/s11270-010-0503-0

Monteiro, C. M., P. M. L. Castro, F. X. Malcata (2010). Cadmium Removal by Two Strains of Desmodesmus pleiomorphus Cells. *Water Air Soil Pollution,* 208, 17–27. https://doi.org/10.1007/s11270-009-0146-1

Nawaz, S., H. N. Bhatti, T. H. Bokhari, S. Sadaf (2014). Removal of Novacron Golden Yellow dye from aqueous solutions by low-cost agricultural waste: Batch and fixed bed study. *Chemistry and Ecology*, 30, 52–65. https://doi.org/10.1080/02757540.2013.841898

Neubaker, U., B. Nowalk, G. Furrer, R. Schulin (2000). Heavy metal sorption on clay minerals affected by the siderophore desferrixamine B. *Environmental Science and Technology*, 34, 2749–2755.

Nowicki1, P. J. Kazmierczak-Raznal, P. Skibiszewska1, M. Wis´niewska, A. Nosal-Wiercin´ska, R. Pietrzak (2016). Production of activated carbons from biodegradable waste materials as an alternative way of their utilisation. *Adsorption,* 22, 489–502. https://doi.org/10.1007/s10450-015-9719-z

Ogunlalu, O., I. P. Oyekunle, O. Kingsley, B. C. lwuozor, A. D. Aderibigbe, E. C. Emenike (2021). Trends in the mitigation of heavy metal ions from aqueous solutions using unmodified and chemically modified agricultural waste adsorbents. *Current Research in Green and Sustainable Chemistry*, 4, 100188.

Önal, E., N. Özbay, A. Ş. Yargıç, R. Z. Y. Şahin, Ö. Gök (2014). Performance Evaluation of the Bio-char Heavy Metal Removal Produced from Tomato Factory Waste, in: Dincer, I., A. Midilli, H. Kucuk (eds), *Progress in Exergy, Energy, and the Environment*. Springer International Publishing, Cham, pp. 733–740. https://doi.org/10.1007/978[[sbn]]3[[sbn]]319[[sbn]]04681[[sbn]]5_70

Öztürk, A., E. Malkoc (2015). Cationic Basic Yellow 2 (BY2) adsorption onto manure ash: Surface properties and adsorption mechanism. Desalin. *Water Treatment*, 54, 209–226. https://doi.org/10.1080/19443994.2013.878671

Pavan, F. A., E. S. Camacho, E. C. Lima, G. L. Dotto, V. T. Branco, S. L. Dias (2014). Formosa papaya seed powder (FPSP): Preparation, characterization and application as an alternative adsorbent for the removal of crystal violet from aqueous phase. *Journal of Environmental Chemical Engineering*, 2(1), 230–238.

Purkait, M. K., A. Maiti, S. DasGupta, S. De (2007). Removal of congo red using activated carbon and its regeneration. *Journal of Hazardous Materials*, 145(1–2), 287–295.

Randhawa, N. S., N. N. Das, R. K. Jana (2014). Adsorptive remediation of Cu(II) and Cd(II) contaminated water using manganese nodule leaching residue. *Desalination Water Treatment*, 52, 4197–4211. https://doi.org/10.1080/19443994.2013.801324

Rangsayatorn, N., E. S. Upatham, M. Kruatrachue, P. Pokethitiyook, G. R. Lanza (2002). Phytoremediation potential of Spirulina (Arthrospira) platensis: Biosorption and toxicity studies of cadmium. *Environmental Pollution*, 119, 45–53. https://doi.org/10.1016/S0269-7491(01)00324-4

Rashid, R., I. Shafiq, P. Akhter, M. J. Iqbal, M. Hussain (2021). A state-of-the-art review on wastewater treatment techniques: The effectiveness of adsorption method. *Environmental Science and Pollution Research*, 28, 9050–9066. https://doi.org/10.1007/s11356-021-12395-x /

Ren, X., W. Xiao, R. Zhang, Y. Shang, R. Han (2013). Adsorption of crystal violet from aqueous solution by chemically modified phoenix tree leaves in batch mode. *Desalination Water Treatment*, 53, 1–11. https://doi.org/10.1080/19443994.2013.859105

Rida, K., S. Bouraoui, S. Hadnine (2013a). Adsorption of methylene blue from aqueous solution by kaolin and zeolite. *Applied Clay Science*, 83–84(Supplement C), 99–105.

Rida, K., S. Bouraoui, S. Hadnine (2013b). Adsorption of methylene blue from aqueous solution by kaolin and zeolite. *Applied Clay Science*, 83–84, 99–105.

Sadaf, S., H. N. Bhatti, S. Ali, K-U. Rehman (2014). Removal of Indosol Turquoise FBL dye from aqueous solution by bagasse, a low-cost agricultural waste: Batch and column study. *Desalination and Water Treatment*, 52(1–3), 184–198.

Saha, R., K. Mukherjee, I. Saha, A. Ghosh, S. K. Ghosh, B. Saha (2013). Removal of hexavalent chromium from water by adsorption on mosambi (Citrus limetta) peel. *Research Chemical Intermediates*, 39, 2245–2257. https://doi.org/10.1007/s11164-012-0754-z

Said, A. E.-A.A.,A. A. M. Aly, M. N. Goda, M. Abd El-Aal, M. Abdelazim (2017). Modified Sugarcane Bagasse with Tartaric Acid for Removal of Diazonium Blue from Aqueous Solutions. *Journal of Polymers and the Environment*, 26, 2424–2433. https://doi.org/10.1007/s10924-017-1136-9

Salleh, M. A. M., D. K. Mahmoud, W. A. Karim, A. Idris (2011). Cationic and anionic dye adsorption by agricultural solid wastes: A comprehensive review. *Desalination*, 280(1–3), 1–13.

Sangi, M. R., A. Shahmoradi, J. Zolgharnein, G. H. Azimi, M. Ghorbandoost (2008). Removal and recovery of heavy metals from aqueous solution using Ulmus carpinifolia and Fraxinus excelsior tree leaves. *Journal Hazardous Materials,* 155, 513–522. https://doi.org/10.1016/j.jhazmat.2007.11.110

Saravanan, R., A. K. Priya, P. S. Kumar, K. Tuan, A. Hoang, K. Sekar (2022). A critical and recent developments on adsorption technique for removal of heavy metals from wastewater-A review. *Chemosphere,* 303(2), 135146.

Schroth, B. K., G. Sposito (1998). Effect of landfill leachate organic acids on trace metal adsorption by kaolinite. *Environmental Science Technology,* 32, 1404–1408.

Seader, J. D., E. J. Henley (1998). *Separation Process Principles,* 1st edition. Wiley ISBN-10:0471586269, ISBN-13: 978–0471586265.

Sen, T. K. (2001). Ph. D. thesis entitled Studies on Colloidal fines-associated contaminant transport in saturated porous media at Indian Institute of Technology, Bombay, India. Call no. 043:541.182:628.54Sen, Barcode 194744. IITB Thesis Access - Central Library Details for: Studies on colloidal fines-associated contaminant transport in porous media (R) › Central Library, IIT, Bombay catalog

Sen, T. K. (2015). *Physical, Chemical and Biological Treatment Processes for Water and Wastewater,* Nova Publishers Inc. 1st edition, ISBN: 978-1-63483-396-7. (https://www.novapublishers.com/catalog/product_info.php?products_id=55622).

Sen, T. K. (2017). *Air, Gas and Water Pollution Control Using Industrial and Agricultural Solid Wastes Adsorbents.* CRC-Press, Taylor & Francis, 1st edition, ISBN: 13:978-1-138-19673-5 (hhttps://www.crcpress.com/Air-Gas-and-Water-Pollution-Control-Using-Industrial-and Agricultural/Sen/p/book/9781138196735).

Sen, T. K. (2023). Agricultural solid wastes based adsorbent materials in the remediation of heavy metal ions from water and wastewater by adsorption: A review. *Molecules,* 28(14), 555. https://doi.org/10.3390/molecules28145575

Sen, T. K., A. F. B. Azman, S. Maitra, B. K. Dutta (2011). Removal of mercury (II) by leaves of Rambai Tree. *Water Environmet Research,* 83(9), 834–842.

Sen, T. K., S. Afroze, H. M. Ang (2011). Equilibrium, kinetics and mechanism of removal of methylene blue from aqueous solution by adsorption onto pine cone biomass of pinus radiata. *Water, Air, Soil Pollution,* 218, 499–515. https://doi.org/10.1007/s11270-010-0663-y

Sen, T. K., D. Gomez (2011). Adsorption of zinc from aqueous solution on natural bentonite. *Desalination,* 267(2–3), 286–294. https://doi.org/10.1016/j.desal.2010.09.041

Sen, T. K., M. Verlyny Sarzali (2008). *Removal of cadmium metal ion (Cd2+) from its aqueous solution by aluminium oxide (Al2 O3),* A kinetic and equilibrium study. *Chemical Engineering Journal,* 142, 256–262.

Senthil Kumar, P., M. Palaniyappan, M. Priyadharshini, A. Vignesh, A. Thanjiappan, P. Sebastina Anne Fernando, R. Tanvir Ahmed, R. Srinath (2014). Adsorption of basic dye onto raw and surface-modified agricultural waste. *Environmental Progress & Sustainable Energy,* 33(1), 87–98.

Sewu, D. D., P. Boakye, H. Jung, H. S. H. Woo (2017a). Synergistic dye adsorption by biochar from co-pyrolysis of spent mushroom substrate and Saccharina japonica. *Bioresource Technology,* 244(Part 1), 1142–1149.

Sewu, D. D., P. Boakye, P. S. H. Woo (2017b). Highly efficient adsorption of cationic dye by biochar produced with Korean cabbage waste. *Bioresource Technology*, 224, 206–213.

Shaikh, R. B., B. Saifullah, F. Rehman (2018). Greener method for the removal of toxic metal ions from the wastewater by application of agricultural waste as an adsorbent. *Water*, 10(10), 1316. https://doi.org/10.3390/w10101316

Sheoran, K., S. S. Siwal, D. Kapoor, N. Singh, A. K. Saini, W. F. Alsanie, V. K. Thakur (2022). Air pollutants removal using biofiltration technique: A challenge at the frontiers of sustainable environment. *ACS Eng.*, 2, 378–396. https://doi.org/10.1021/acsengineeringau.2c00020?urlappend=%3Fref%3DPDF&jav=VoRrel=cite-as

Slejko, F. L. (1985). Adsorption technology: A step-by-step approach to process evaluation and application. Book, 1st edition, M. Dekker publishers, NewYork. ISBN: 9780824772857, 0824772857. OCLC No/Unique Identifier: 11496959

Sočo, E., J. Kalembkiewicz (2013). Adsorption of nickel(II) and copper(II) ions from aqueous solution by coal fly ash. *Journal of Environmental Chemical Engineering*, 1(3), 581–588.

Stavropoulos, G., G. Skodras, K. Papadimitriou (2015). Effect of solution chemistry on cyanide adsorption in activated carbon. *Applied Thermal Engineering*, 74, 182–185.

Sulyman, M., J. Namiesnik, A. Gierak (2017). A low-cost adsorbents derived from agricultural by-products/wastes for enhancing contaminant uptakes from wastewater: A review. *Polish Journal Environmental Studies*, 26(2), 479–510 Review. https://doi.org/10.15244/pjoes/66769

Tamura, H., R. Furuichi (1997). Adsorption affinity of divalent heavy metal ions for metal oxides evaluated by modelling with the Frumkin isotherm. *Journal of Colloid and Interface Science*, 195, 241–249.

Tanweer, A., M. Danish (2022). A review of avocado waste-derived adsorbents: Characterizations, adsorption characteristics, and surface mechanism. *Chemosphere*, 296, 134036. https://doi.org/10.1016/j.chemosphere.2022.134036. 2020.

Tee, G. T., W. F. Gok, Y. Xie, W. F. Yong (2022). Review article Adsorption of pollutants inwastewater via biosorbents, nanoparticles and magnetic biosorbents: A review. *Environ. Res.*, 212(B), 113248. https://doi.org/10.1016/j.envres.2022.113248

Tien, C. (1994). *Adsorption Calculation and Modelling*, 1st edition. Butterworth-Heinemann ISBN 0-7506-9121-2

Tina Co Nnie Tan, T. K. Sen (2019). Thermodynamics, equilibrium and kinetic studies of aqueous phase methylene blue (MB) dye adsorption onto mixture of eucalyptus bark biomass and kaolin clay minerals. *Separation Science and Technology*, 55, 1–15. https://doi.org/10.1080/01496395.2019.1580734

Trivedi, P., L. Axe (2000). Modelling Cd and Zn sorption to hydrous metal oxides. *Environmental Science and Technology*, 34, 2215–2223.

Tuzen, M., A. Sari, D. Mendil, O. D. Uluozlu, M. Soylak, M. Dogan (2009). Characterization of biosorption process of As(III) on green algae Ulothrix cylindricum. *Journal of Hazardous Materials*, 165, 566–572. https://doi.org/10.1016/j.jhazmat.2008.10.020

Uçar, S., M. Erdem, T. Tay, S. Karagöz (2015). Removal of lead (II) and nickel (II) ions from aqueous solution using activated carbon prepared from rapeseed oil cake by Na2CO3 activation. *Clean Technologies and Environmental Policy*, 17(3), 747–756.

Uddin, M. T., M. A. Rahman, M. Rukanuzzaman, M. A. Islam (2017). A potential low cost adsorbent for the removal of cationic dyes from aqueous solutions. *Applied Water Science*, 7(6), 2831–2842.

Vakili, M., S. Deng, G. Cagnetta, W. Wang, P. Meng, D. Liu, G. Yu (2019). Regeneration of chitosan-based adsorbents used in heavy metal adsorption: A review, *Separtion Purification Technology,* 224, 373–387. https://doi.org/10.1016/j.seppur.2019.05.040

Vulava, V. M., R. Kretzschmar, U. Rusch, D. Grolimund, J. C. Westall, M. Borkovec (2000). Cation competition in a natural subsurface material: Modeling of sorption equilibria. *Environmental Science and Technology,* 34, 2149–2155.

Wang, H., X. Yuan, Z. Wu, L. Wang, X. Peng, L. Leng, G. Zeng (2014). Removal of basic dye from aqueous solution using Cinnamomum camphora sawdust: Kinetics, isotherms, thermodynamics, and mass-transfer processes. *Seperation Science and Technology,* 49, 2689–2699. https://doi.org/10.1080/01496395.2014.940590

Wei, M., Q. Yu, T. Mu, L. Hou, Z. Zuo, J. Peng (2016). Preparation and characterization of waste ion-exchange resinbased activated carbon for CO2 capture. *Adsorption,* 22, 385–396. https://doi.org/10.1007/s10450-016-9787-8

Yagub, Mustafa, T. K. Sen, S. Afroza, H. A. Ming (2014). Dye and its removal from aqueous solution by adsorption: A review *–Advances in Colloids and Interface Science,* 209, 172–184.

Yargic, A. S., R. Z. Yarbay, N. Sahin, O. E. Onal (2015). Assessment of toxic copper(II) biosorption from aqueous solution by chemically-treated tomato waste. *Journal of Cleaner* Production, 88, 152–159. https://doi.org/10.1016/j.jclepro.2014.05.087

Yildiz, H. (2024). The production of a novel adsorbent from forest waste (Platanus orientalis L.) for dye adsorption: Adsorption process optimization and experimental design. *Materials Science & Engineering B,* 304, 117366. https://doi.org/10.1016/j.mseb.2024.117366

Zazycki, M. A., M. Godinho, D. Perondi, E, L. Foletto, G. C. Collazzo, G. L. Dotto (2018). New biochar from pecan nutshells as an alternative adsorbent for removing reactive red 141 from aqueous solutions. *Journal of Cleaner Production,* 171(Supplement C), 57–65.

Zhang, Z., I. M. O'Hara, G. A. Kent, W. O. S. Doherty (2013). Comparative study on adsorption of two cationic dyes by milled sugarcane bagasse. *Industrial Crops Products,* 42, 41–49. https://doi.org/10.1016/j.indcrop.2012.05.008

Zhao, B., Y. Shang, W. Xiao, C. Dou, R. Han (2014). Adsorption of Congo red from solution using cationic surfactant modified wheat straw in column model. *Journal Environmental Chemical Engineering,* 2(1), 40–45. https://doi.org/10.1016/j.jece.2013.11.025

Zhao, Z., B. Wang, B. K. G. Theng, X. Lee, X. Zhang, M. Chen, P. Xu (2022). Removal performance, mechanisms and influencing factors of biochar for air pollutants: A critical review. *Biochar,* 4, 30. https://doi.org/10.1007/s42773-022-00156-z

Zhou, N., H. Chen, J. Xi, D. Yao, Z. Zhou, Y. Tian, X. Lu (2017). Biochars with excellent Pb(II) adsorption property produced from fresh and dehydrated banana peels via hydrothermal carbonization. *Bioresource Technology,* 232, 204–210.

Zhu, L., Y. Wang, T. He, L. You, X. Shen, X. (2016). Assessment of potential capability of water bamboo leaves on the adsorption removal efficiency of cationic dye from aqueous solutions. *Journal of Polymers and the Environment,* 24(2), 148–158.

Zwain, H. M., M. Vakili, I. Dahlan (2014). Waste material adsorbents for zinc removal from wastewater: A comprehensive review. *International Journal of Chemical Engineering,* 1–13. https://doi.org/10.1155/2014/347912

2

Common Adsorbents, Synthesis Methods, Characterization and Its Application in Water and Air Pollutants Removal by Solid/Liquid Adsorption Process

2.1 Introduction

The adsorption method is widely favoured for its simplicity and the broad range of available adsorbents, making it an effective process for removing refractory pollutants from wastewater and air (Crini, 2006; Gupta, 2009; Sen, 2023). Characterizing the physical and chemical surface properties of adsorbents is essential for predicting adsorption capacity (Afroza & Sen, 2018). Key adsorbent characteristics, such as specific surface area, BET surface area, particle size, pore volume, point of zero charge (pH_{pzc}), bulk density, and surface functional groups, significantly influence adsorbent-adsorbate interactions and, consequently, adsorption capacity (Dawood & Sen, 2014; Yagub et al., 2014). The presence of functional groups, such as hydroxyl, nitro, and azo groups, enhances adsorbent selectivity and pollutant sorption from both aqueous and gaseous phases. Adsorption rates increase with the surface area, shape, and surface charge of adsorbents (Sen, 2023a). Key characteristics to consider when selecting an adsorbent include (Geankoplis, 2003; Seader, 1998):

- High selectivity for precise separation
- High capacity to minimize adsorbent usage
- Favourable kinetics and transport properties for rapid sorption
- Chemical and thermal stability
- Hardness and mechanical strength
- High fouling resistance
- Low solubility in the carrier solvent
- Cost-effectiveness and regeneration capability.

DOI: 10.1201/9781003373223-2

Traditional commercial adsorbents, such as activated carbon, zeolites, silica gel, and activated alumina, have been highly effective in adsorbing inorganic and organic compounds from aqueous phases. However, their high costs have driven research toward developing full-scale adsorption processes using natural or treated solid wastes and solid waste-based synthetic adsorbents (Dutta, 2007). Four primary commercial adsorbents include (a) activated carbon/charcoal, (b) zeolites, (c) silica gel, and (d) activated alumina. New-generation adsorbents, such as carbon nanotubes (CNTs), carbon-based magnetic nanocomposites, and other nanomaterials, are also increasingly used for water and air pollutant removal by adsorption (Sen, 2023b).

Adsorbents used in solid/liquid and gas/solid adsorption processes can be classified into natural and synthetic categories (EI-Baz et al., 2020). Natural adsorbents include materials such as charcoal, clay minerals, chitin, zeolite, and ores, while synthetic adsorbents are derived from agricultural products, marine and forest waste, household and industrial waste, sewage sludge, and polymers (EI-Baz et al., 2020). Coal-based activated carbon (CAC) is well-known for its high specific surface area, porous nature, and functional groups such as carboxyl, carbonyl, phenolic, and amine, which facilitate metal binding and surface complex formation (Deliyanni et al., 2015; Fu & Wang, 2011; Sen, 2023a). However, CAC's high cost and limited regeneration capacity have shifted research focus toward cost-effective alternatives, such as solid waste-based adsorbents. Lignocellulosic, carbonaceous, agricultural, and forest-based alternative adsorbents are now extensively explored for water decontamination, including metal removal (Sen, 2023a).

Based on precursor materials, adsorbents are categorized as agricultural waste-based, aquatic and terrestrial biomass-based, local waste biomass, zeolites, clay materials, industrial solid waste, fullerene, and CNTs (Burakov et al., 2018). Recent studies highlight the use of various natural, agricultural, and industrial waste materials and biosorbents synthesized from microorganisms (e.g., algae, bacteria) for pollutant removal (Bayuo et al., 2023). Bayuo et al. (2023) also reviewed the simultaneous removal of heavy metals from multicomponent systems using diverse sorbent materials, including activated carbon, biochar, composites, zeolites, graphene, agro waste, nanoparticles, and biosorbents.

Readers are encouraged to consult recent review articles on alternative solid waste-based adsorbents for heavy metal ion removal from aqueous phases (Afroza & Sen, 2018; Ahmad et al., 2017; Bayuo et al., 2023; Gisi et al., 2016; Liu & Zhang, 2022; Ogunlalu et al., 2021; Soliman & Moustafa, 2020) and on air pollutant removal by fly ash, biochar, and related materials (Ge et al., 2018; Sen, 2017; Zhao, 2022). The following sections discuss various types of adsorbents, their synthesis and characterization, and their applications in water and air pollutant adsorption.

2.2 Common Adsorbents, Synthesis Methods, and Major Characterization for Adsorption

2.2.1 Activated Carbon

Activated carbon is one of the most widely used adsorbents, closely resembling common charcoal. This amorphous, highly porous carbonaceous material is renowned for its versatile surface properties, including an extensive inter-particulate surface area and small hydrophobic graphite layers. These properties make activated carbon highly effective in adsorption and catalytic applications (Bansal & Goyal, 2005; Kazmierczak-Razna et al., 2015). The high porosity of activated carbon enhances its efficiency (Sharma & Bhattacharya, 2017). Various carbonaceous raw materials, such as wood, coal char, almond, coconut, and walnut shells, are commonly used in its production (Dutta, 2007).

Commercial activated carbon is typically synthesized through partial oxidation of coal, with the activation process conducted by either physical or chemical treatment (Afroze & Sen, 2018; Dawood & Sen, 2012). This involves mixing raw materials with alkali, acid, phosphate, or zinc chloride and then calcining them at temperatures between 500 and 900°C. Figure 2.1 illustrates the general characteristics of granular and powdered activated carbon (Geankoplis, 2003; Seader & Henley, 1998):

- Char-like material with high surface area (400–1200 m^2/g)
- Pore size: 1–6 nm
- Average particle size: 15–25 microns
- Amorphous and hydrophobic

FIGURE 2.1
Granular and powdered activated carbon

- Suitable for removing trace inorganics and organics
- Commonly used in air and water filtration systems

2.2.2 Silica Gel

Silica gel, composed of silicon dioxide (SiO_2), is an amorphous, hard, glass-like granular or beaded material, as shown in Figure 2.2. While pure silica is chemically inert and non-polar, the presence of hydroxyl (silanol) groups on its surface makes it polar and hydrophilic (Dutta, 2007). Silica gel is typically formed through two methods: (1) polymerizing silicic acid or (2) aggregating colloidal silica particles (Sharma & Bhattacharya, 2017).

This non-toxic and non-corrosive material exhibits high adsorption capacity due to its extensive surface area and porosity. Commonly prepared by precipitation from sodium silicate solutions treated with acid, silica gel has the following properties:

- Pore size: 2–5 nm
- Amorphous and hydrophilic
- Suitable for water and moisture removal, gas and liquid dehydration, and as a desiccant
- Modified forms are used in water purification and hydrocarbon fractionation

FIGURE 2.2
Silica gel crystal beads

2.2.3 Aluminosilicate (Molecular Sieves)/Zeolites

Zeolites are aluminosilicates characterized by a silicon-to-aluminium (Si/Al) ratio ranging from 1 to infinity. They possess a tetrahedral network of silicon and oxygen atoms, where some silicon atoms are substituted by aluminium, forming alumina-silicates (Sharma & Bhattacharya, 2017). Primarily used in separation processes, synthetic zeolites are favoured in scientific and engineering applications over natural zeolites due to their high purity, uniform size, and enhanced reactivity. Commercially, synthetic zeolites are often produced using sodium aluminosilicate gel via hydrothermal treatment. Figure 2.3 illustrates the common properties of aluminosilicate molecular sieves (Geankoplis, 2003; Seader, 1998):

- Hydrophilic
- Surface area: 600–700 m^2/g
- Highly structured, porous space
- Utilized for N_2 removal from air, drying operations, hydrocarbon separation
- Effective in NH_4 and heavy metal removal as well as organic and humic substance (e.g., humic, fulvic acid) removal

2.2.4 Activated Alumina

Activated alumina, primarily composed of aluminium oxide (Al_2O_2) in spherical bead form, is highly porous and exhibits a significant surface area (Sharma & Bhattacharya, 2017). It exists in three forms: activated alumina sorbent, desiccant, and catalyst carrier. Synthesized from precipitated aluminium hydroxide gel, activated alumina is prepared through ammonization of an aluminium salt or by acidifying a sodium aluminate solution (Dutta, 2007). The gel is then dried and processed into granules or spheres. Key properties include (Seader, 1998):

- Hydrated aluminium oxide activated by heating and drying
- Pore size: 2–14 nm
- Fractional porosity: 0.4
- Surface area: 200–500 m^2/g
- Bulk density: 700–900 kg/m^3
- Applications include dehydrating gases and liquids, drying air, hydrocarbons, and SO_2, and removing contaminants such as As^{5+}, PO_4^{3-}, Cl^-, F^-, Se, Sb, Pb, and Bi from water

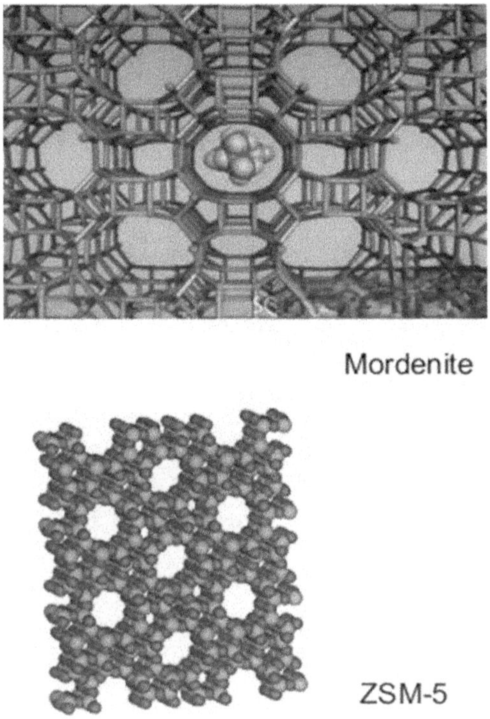

FIGURE 2.3
Molecular sieves

2.2.5 Various Raw and Modified Agricultural Solid Waste-Based Adsorbents

Agricultural solid wastes, including leaves, bark, fibres, seeds, fruit peels, and sawdust, have been recognized as effective adsorbents for removing organic dyes, aromatics, and inorganic heavy metals from wastewater (Dawood, 2018). Figures 2.4 and 2.5 display some examples, such as date palm tree waste, castor hull, rambutan fruit shell, Jatropha leaves, pine tree leaves, pinecones, and eucalyptus bark (EB). These materials often contain high-molecular weight components like lignin, cellulose, and hemicellulose (Salleh et al., 2011), making them biodegradable and renewable. The availability of these materials in large quantities with minimal economic value also helps in waste disposal management, as they can be converted into low-cost, effective adsorbents means added value to the wastes.

Recent studies highlight agricultural waste as a promising adsorbent for removing inorganic and organic pollutants (Sen, 2023). Figures 2.4 and 2.5 show examples of these materials, which include eucalyptus tree waste,

FIGURE 2.4
Raw eucalyptus tree solid waste, pinecone, pine tree needle leaves, castor hull biomass, and dates palm tree solid waste adsorbents

FIGURE 2.5
Raw eucalyptus bark material, rambutan fruit shell, pinecone, and Jatropha tree solid waste adsorbents

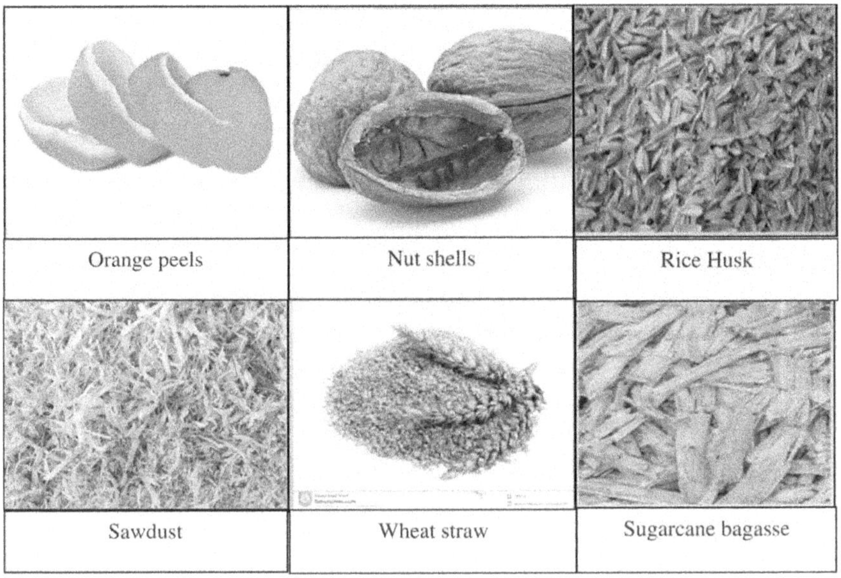

FIGURE 2.6
Examples of agricultural biomass-based adsorbents used in heavy metal ion adsorption.

Source: Chakrabortya et al., 2022; Sen, 2023a; with permission

pinecones, castor hull, and dates palm tree waste. Additionally, agricultural residues and fruit peels, unsuitable for other uses, can be pre-treated or processed into low-cost activated carbon adsorbents (Rashid et al., 2021; Sen, 2023a). Numerous review articles, including those by Sen (2023a), Ahmed and Danesh (2022), Saukat et al. (2022), Ogunlalu et al. (2021), Afroze and Sen (2018), and Sulyman et al. (2017), provide extensive information on heavy metal removal using agricultural waste-derived adsorbents. Figure 2.6 shows additional examples of agricultural by-products that serve as cost-effective adsorbents for heavy metal ion removal (Sen, 2023a).

2.2.5.1 *Synthesis, Morphology, and Surface Properties of Selective Agricultural Solid Wastes-Based Adsorbents as Alternatives to Costly Commercial Activated Carbon for Inorganic/Organic Adsorption at the Solid/Liquid Interface*

2.2.5.1.1 *Synthesis and Characterization of EB and Pinecone-Based Biochar*

Eucalyptus sheathiana, from the *Myrtaceae* family, is an evergreen, fast-growing tree with distinctive bark that often hangs in long ribbons. Similarly, pine trees are globally available, particularly in regions like Australia. *Eucalyptus sheathiana* bark (EB) and pinecone (*P. radiata*) biomasses were collected from the Curtin University Bentley campus in

Western Australia between January and April 2014, as shown in Figure 2.5 (Afroze, 2016). The collected biomass was cleaned multiple times with deionized water and then oven-dried at 105°C overnight. Biochar was synthesized from the dried biomass powders using slow pyrolysis following the method by Dawood (2018) and Dawood et al. (2014). The dried biomass was cut into small pieces and placed in closed crucibles within a muffle furnace under limited oxygen and atmospheric pressure. The temperature was gradually increased at a rate of 10°C/min to reach set points of 300°C, 400°C, 500°C, 600°C, and 700°C, maintained for up to 2.5 hours (Dawood et al., 2014, 2018). The resulting biochar was allowed to cool to room temperature over a five-hour period, then ground using a mechanical grinder (RETSCH GmbH & Co. KG, West Germany). The biochar powders were sieved to a size of 106 μm using a British Standard Sieve (BSS) and stored in labelled containers for characterization and adsorption experiments (Afroze, 2016; Dawood, 2018).

2.2.5.1.2 Chemical Modification of EB and Pinecone Biomass

To prepare base-modified EB powder, 50 g of raw EB powder was mixed with 500 mL of 0.1 M sodium hydroxide (NaOH) solution. The mixture was stirred overnight using a magnetic stirrer, then filtered and repeatedly rinsed with distilled water. The washed powder was oven dried overnight at 90°C and subsequently used in adsorption experiments. Figure 2.7 illustrates the preparation of base-modified EB powder (Afroze, 2016).

Similarly, Dawood and Sen (2012) synthesized acid-modified pinecone powder by combining 10 g of raw pinecone powder with 100 mL of 0.1 M HCl solution. The reaction mixture was stirred with a magnetic stirrer for 20 hours, followed by filtration and multiple washes with distilled water. The cleaned powder was then dried overnight at 70°C for use in adsorption and characterization studies.

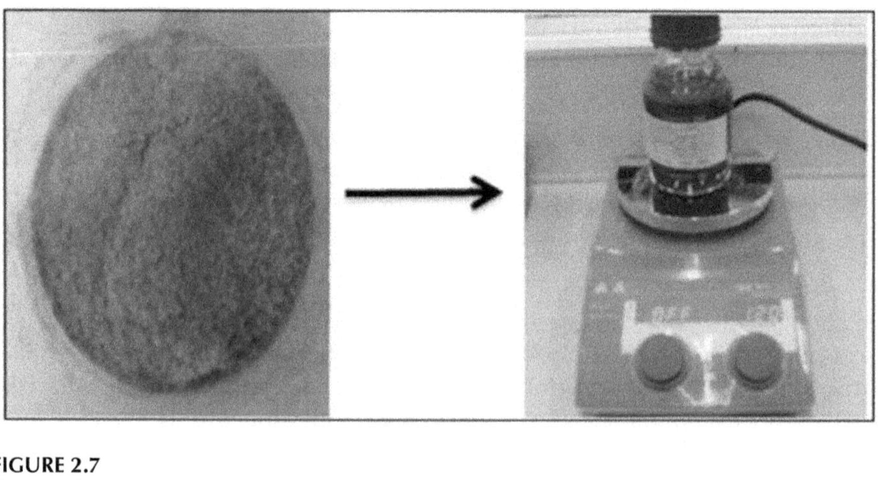

FIGURE 2.7
Activated raw eucalyptus bark powder adsorbent by base treatment

2.2.5.1.3 *Synthesis of Ca-Alginate Sugarcane Bagasse Biochar (CASB) and Ca-Alginate Pinecone Biochar (CAPB) Composite Adsorbents*

The synthesis process, as described by Biswas et al. (2019a, 2019c, 2020c), involves two main steps: (i) preparing sugarcane bagasse biochar through biomass pyrolysis at 500°C in an inert environment for two hours and (ii) combining the synthesized biochar with Na-alginate in a 1:4 ratio. This mixture was processed into Ca-alginate sugarcane bagasse biochar (CASB) composite adsorbent by phase inversion in a 3% (w/v) chilled $CaCl_2$ solution, as shown in Figure 2.8 (Biswas et al., 2020b).

Similarly, Biswas et al. (2019c) synthesized hybrid Ca-alginate pinecone biochar (CAPB) adsorbents using a similar process, as depicted in Figure 2.9, and reported its effectiveness in Zn^{2+} removal from aqueous solutions.

2.2.5.1.4 *Synthesis of Magnetic Bamboo Charcoal–Iron Oxide "BC/Fe" Nanocomposite Adsorbents*

Magnetic "BC/Fe" bamboo charcoal–iron oxide nanocomposites were synthesized from locally sourced bamboo (*Phyllostachys pubescens*) using the solvothermal process, as outlined by Nishioka and Sen (2019). First, bamboo charcoal (BC) was prepared by pyrolysis of bamboo biomass at 800°C for three hours. In the second step, magnetic nanoparticles were introduced into the carbon matrix using ferric nitrate as the iron precursor and ethylene glycol as the reducing agent. The solvothermal reaction was conducted in a sealed Teflon-coated autoclave at 180°C for two hours. After cooling, the

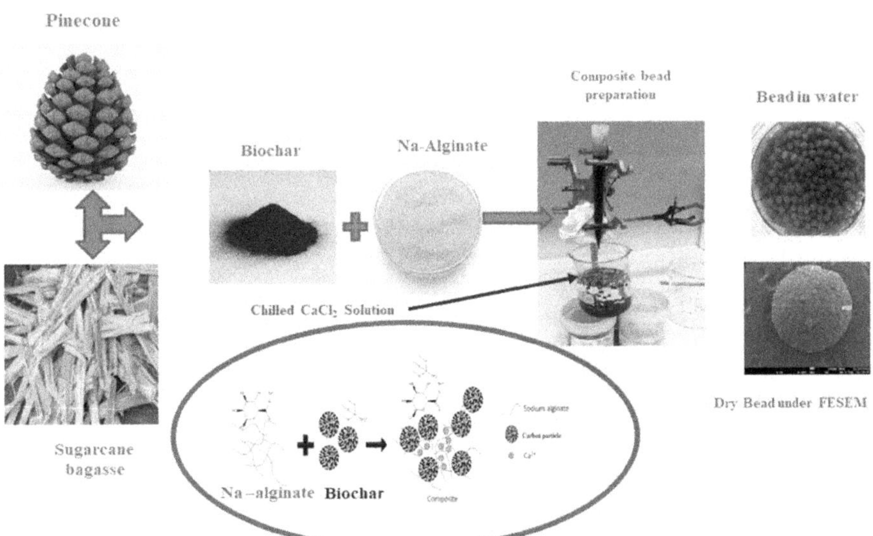

FIGURE 2.8
Synthesis of Ca-alginate sugarcane bagasse biochar (CASB) composite adsorbent

Source: Biswas et al., 2020b, with permission.

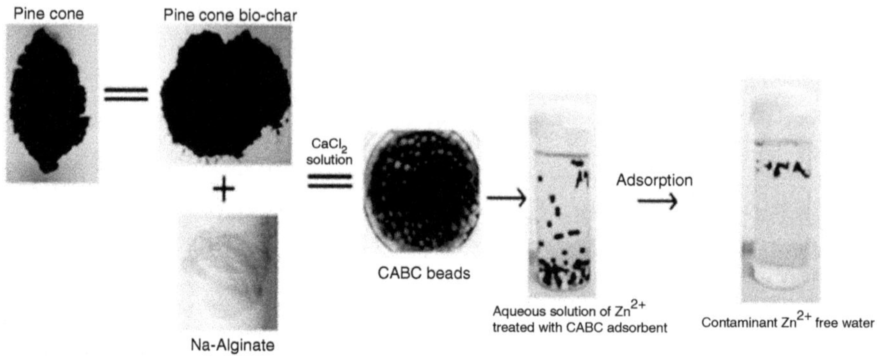

FIGURE 2.9
Synthesis of Ca-alginate pinecone biochar (CAPB) composite adsorbent

Source: Biswas et al., 2019c, with permission.

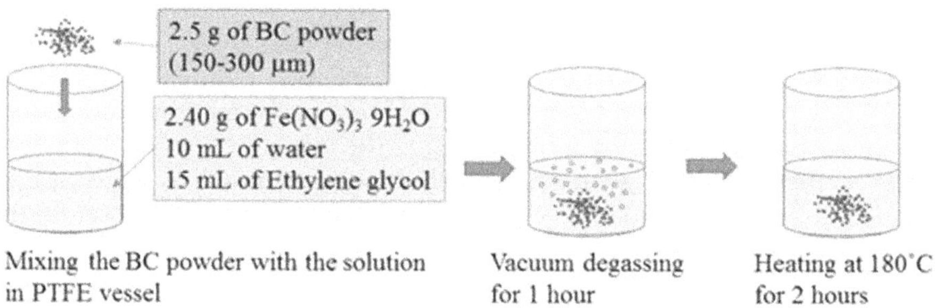

FIGURE 2.10
Solvothermal synthesis of magnetic bamboo charcoal–iron oxide "BC/Fe" nanocomposite adsorbents

Source: Nishioka & Sen, 2019, with permission.

mixture was filtered, and the product was oven-dried at 70°C for 12 hours. The synthesis method is shown in Figure 2.10.

2.2.5.2 Morphology and Surface Properties of Clay Minerals and Selected Agricultural Solid Waste-Based Adsorbents during Adsorption

2.2.5.2.1 *Scanning Electron Microscopy (SEM) and Energy Dispersive X-Ray Spectroscopy (EDX) Studies of Adsorbents*

The morphology, porosity, and texture of adsorbent samples were analysed using scanning electron microscopy (SEM) and energy dispersive X-ray spectroscopy (EDX). Figure 2.11 illustrates SEM and EDX images of kaolin clay before and after methylene blue (MB) dye adsorption. As shown in Figures 2.11 (a) and (b), kaolin particles exhibit clustered, irregular shapes and a porous texture, which enhance MB dye adsorption on the kaolin

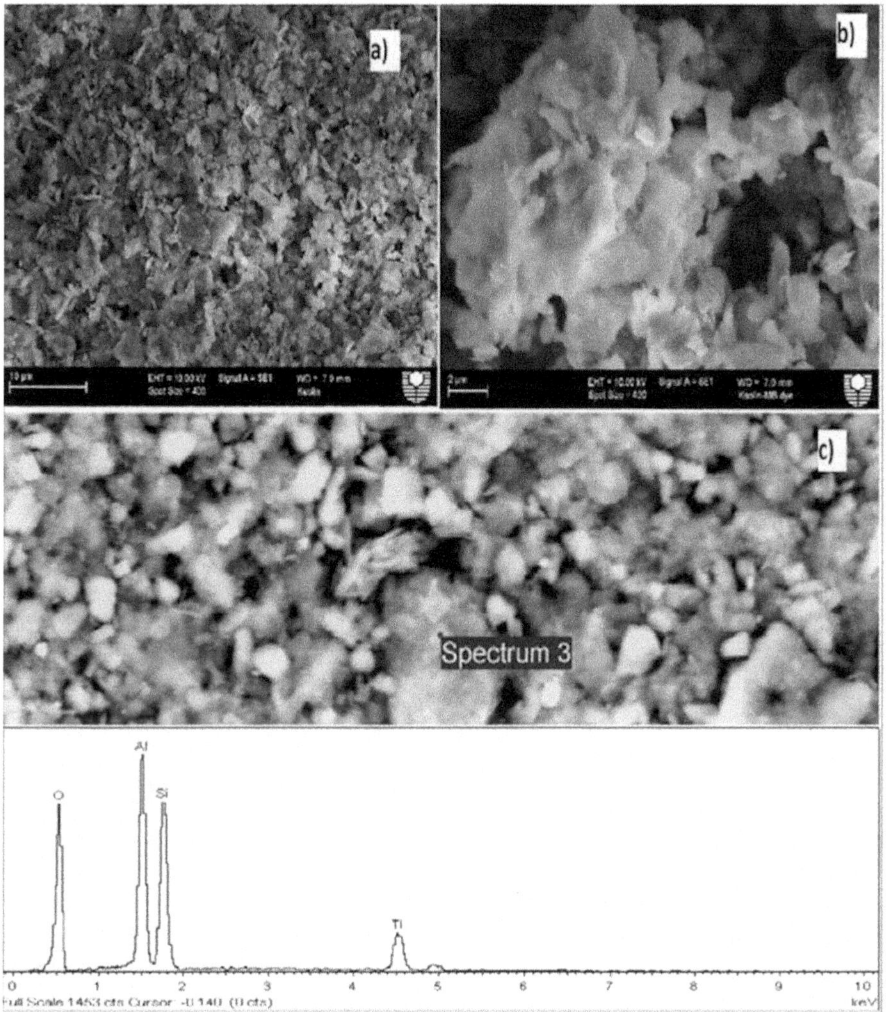

FIGURE 2.11
(a) SEM of kaolin clay before adsorption, (b) SEM of kaolin after MB dye adsorption, (c) EDX of kaolin clay

Source: Dawood et al., 2017a; Dawood, 2018, with permission.

surface (Dawood et al., 2017a; Dawood, 2018). EDX analysis (Figure 2.11 (c)) reveals the primary presence of aluminium, silicon, oxygen, and titanium in kaolin clay minerals (Dawood et al., 2017a; Dawood, 2018).

Similarly, Dawood (2018) reported SEM/EDX analyses of EB biochar before and after MB dye adsorption, as shown in Figure 2.12. EB biochar displays a well-developed, heterogeneous structure with high porosity. Post-adsorption (Figure 2.12 (b)), the reduced porosity indicates dye molecule binding on the

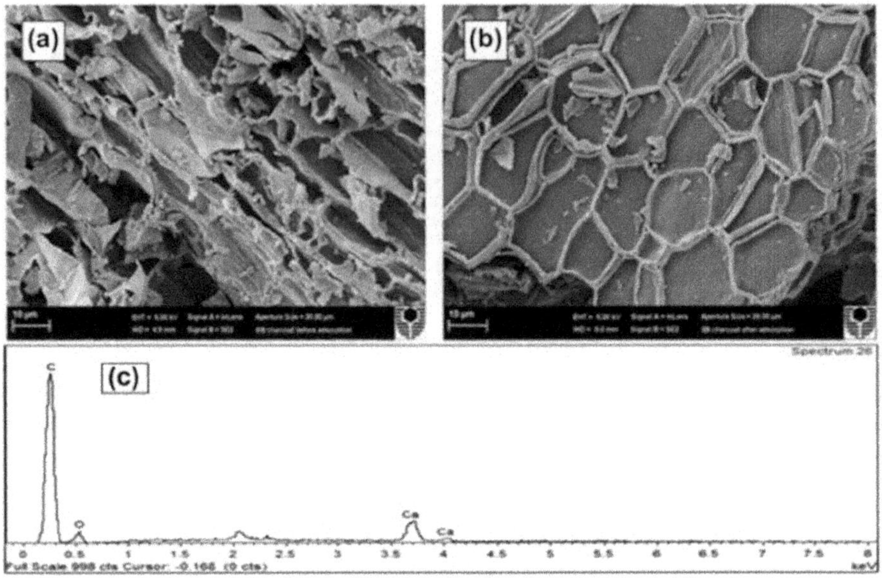

FIGURE 2.12
(a) SEM of EB biochar before adsorption, (b) SEM of EB biochar after MB dye adsorption, (c) EDX of EB biochar

Source: Dawood, 2018, with permission.

biochar surface. EDX analysis (Figure 2.12 (c)) confirms the presence of carbon, oxygen, and calcium (Dawood, 2018).

Afroza et al. (2016b) observed the SEM images of raw and NaOH-treated EB adsorbents used for Zn^{2+} and MB dye adsorption, as shown in Figures 2.13 (a)–(b). SEM images illustrate highly irregular sizes and shapes, indicating that EB is amorphous carbon with a non-crystalline structure. NaOH treatment significantly alters the physicochemical properties and porosity of EB, increasing the irregularity and surface area by exposing more active sites (Afroze, 2016).

Elemental composition analysis, crucial for adsorbent characterization, was performed using electron diffraction spectrum (EDX). Figure 2.14 shows the EDX spectra of raw and NaOH-treated EB, indicating a homogeneous distribution of elements like calcium, oxygen, hydrogen, nitrogen, sodium, magnesium, and chloride across the adsorbent surface (Afroze, 2016, Afroza et al., 2016b; Afroze & Sen, 2018).

Nishioka and Sen (2019) provided SEM/EDX images of magnetic bamboo charcoal-iron oxide (BC/Fe) nanocomposite adsorbents (Figure 2.15). SEM analysis shows porous BC in Figure 2.15 (a), with iron oxide particles distributed on the surface and aggregated at the edges in Figure 2.15 (b). The EDX spectrum (Figure 2.16) confirms high purity, showing only Fe, O, and C in the BC/Fe nanocomposites (Nishioka & Sen, 2019).

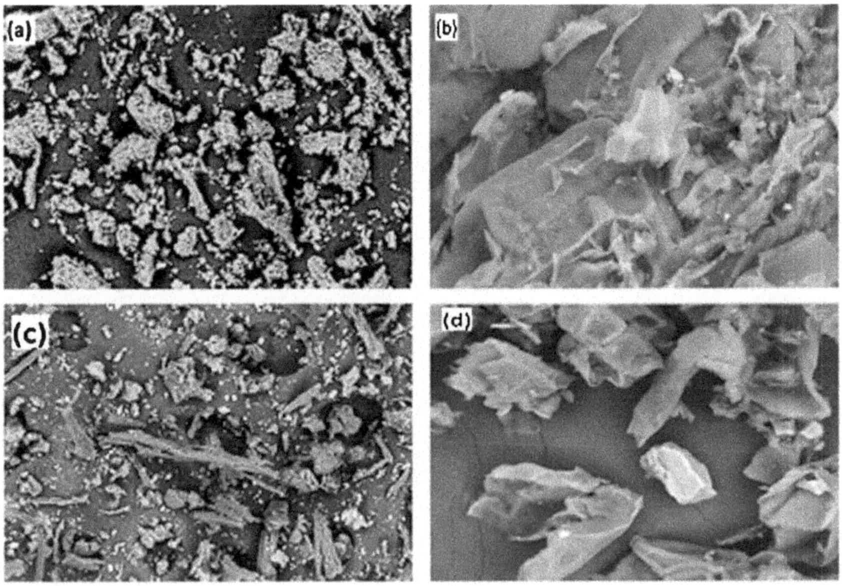

FIGURE 2.13
SEM images of (a) raw EB before adsorption, (b) raw EB after Zn^{2+} adsorption, (c) NaOH-treated EB before adsorption, (d) NaOH-treated EB after Zn^{2+} adsorption

Source: Afroze et al., 2016b, with permission.

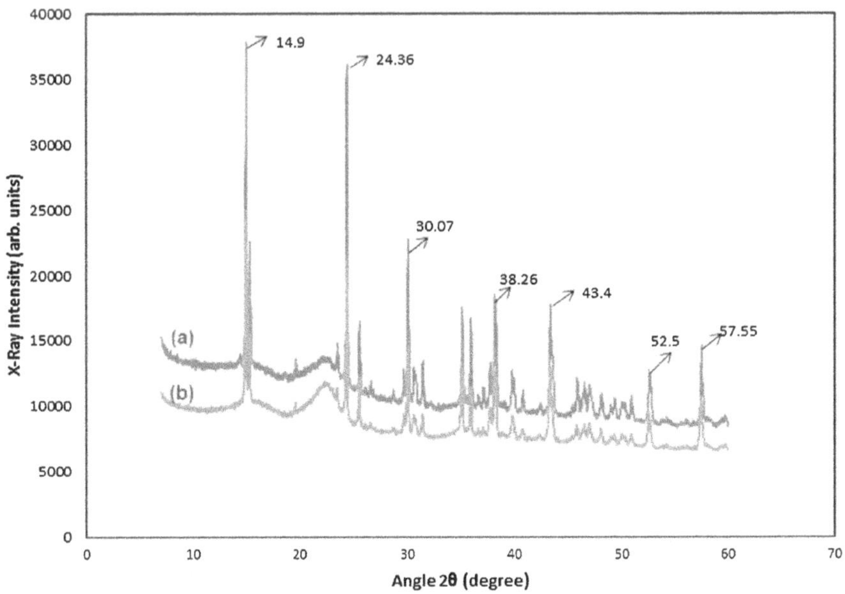

FIGURE 2.14
EDX spectra of (a) raw and (b) NaOH-treated eucalyptus bark powder (106 μm)

Source: Afroze, 2016, Afroze et al., 2016b, with permission.

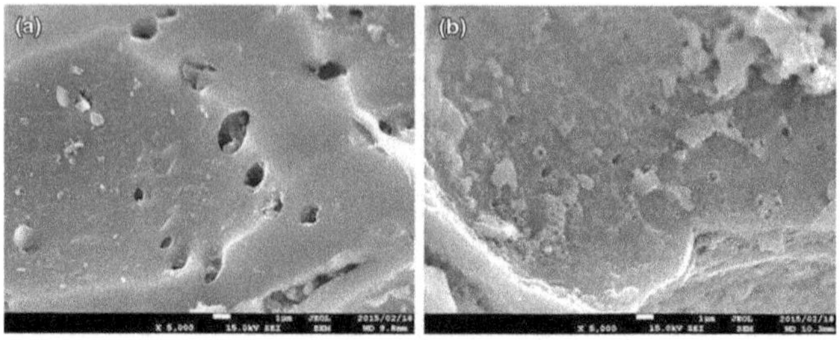

FIGURE 2.15
SEM images of (a) bamboo charcoal (BC) and (b) magnetic bamboo-charcoal nanocomposite (BC/Fe)

Source: Nishioka & Sen, 2019, with permission.

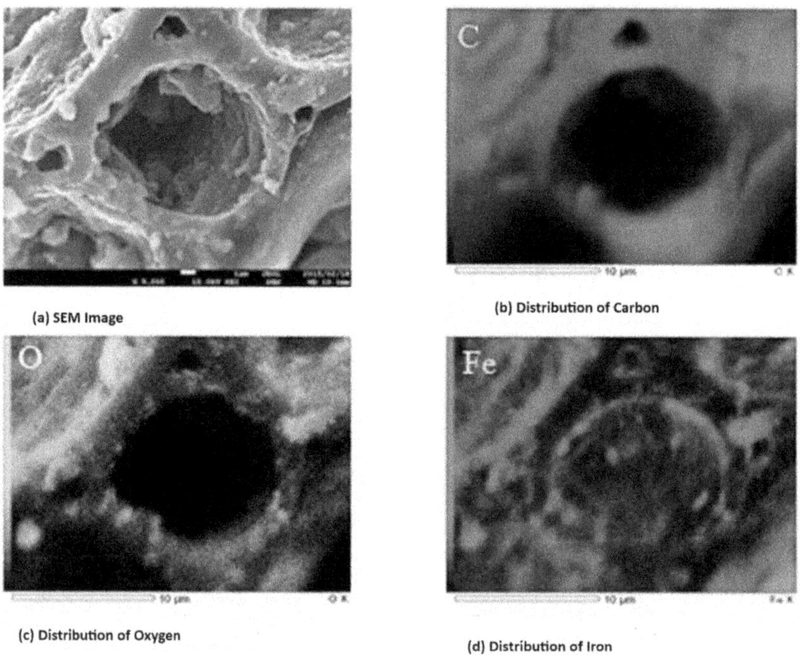

FIGURE 2.16
SEM and EDX images of magnetic bamboo charcoal-iron oxide (BC/Fe) nanocomposites

Source: Nishioka & Sen, 2019, with permission.

Biswas et al. (2020c) analysed sugarcane bagasse (SB) biomass before and after pyrolysis and alginate entrapment, used for MB dye adsorption. The FESEM images (Figure 2.17 (a)–(d)) show increased porosity after pyrolysis (Figure 2.17 (b)) and a highly porous structure post-entrapment in alginate (Figure 2.17 (c)). Adsorption of dye molecules is evident from surface agglomeration in Figure 2.17 (d) (Biswas et al., 2020c).

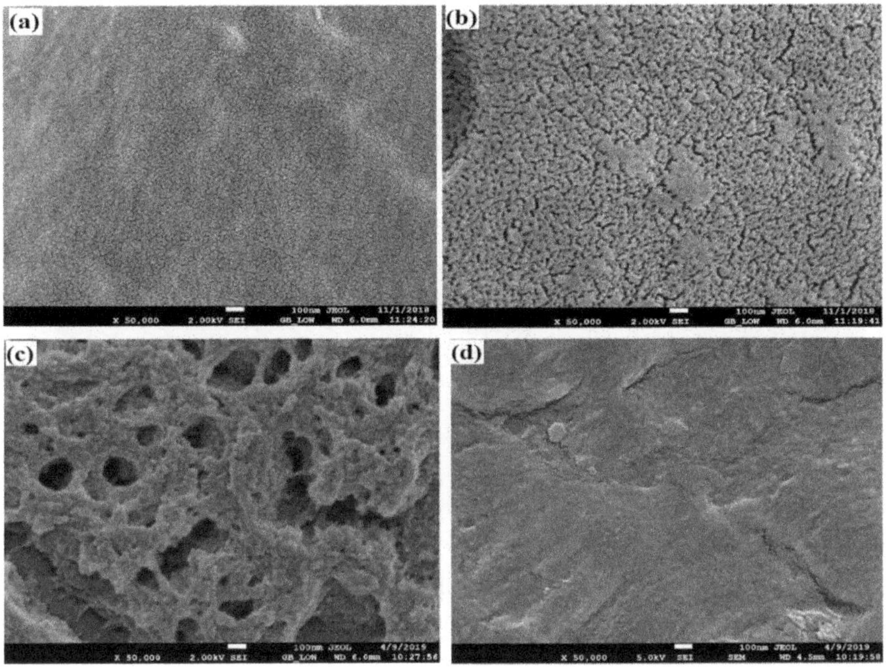

FIGURE 2.17
FESEM images of SB (a) before pyrolysis, (b) after pyrolysis, (c) CASB before treatment, (d) CASB after treatment

Source: Biswas et al., 2020c, with permission.

Al-Yaari and Saleh (2023) synthesized polyethyleneimine-grafted graphene oxide (PEI/GO) adsorbent for Pb^{2+} removal. The SEM images in Figure 2.18 (a)–(b) show graphene oxide (GO) as nano sheets with a smooth surface, while PEI/GO exhibits polymer branches on the GO, indicating substantial morphological differences due to grafting (Al-Yaari & Saleh, 2023).

2.2.5.2.2 X-Ray Diffraction (XRD) Spectrum Study

Powder X-ray diffraction (XRD) is an essential technique for analysing the crystalline state of adsorbent materials, allowing identification of crystalline phases, lattice spacing, and the ordering of crystallites. Sen (2023b) studied MB dye adsorption using magnetic bamboo charcoal–iron oxide nanocomposite (BC/Fe) adsorbents synthesized via a solvothermal method. The XRD patterns, shown in Figure 2.19, were obtained using a Cu Kα radiation source and analysed with Joint Committee on Powder Diffraction Standards (JCPDS) files. Using Scherrer's equation, nano-sized iron oxide adsorbent particles (13.1–15.6 nm) were produced, with the peaks at 2θ values such as 18.48°, 18.89°, 22.92°, and 29.76° confirming the presence of humboldtine (H) and peaks at 30.11°, 35.46°, 43.45°, and 58.76° confirming magnetite (Fe_3O_4) (Nishioka & Sen, 2019).

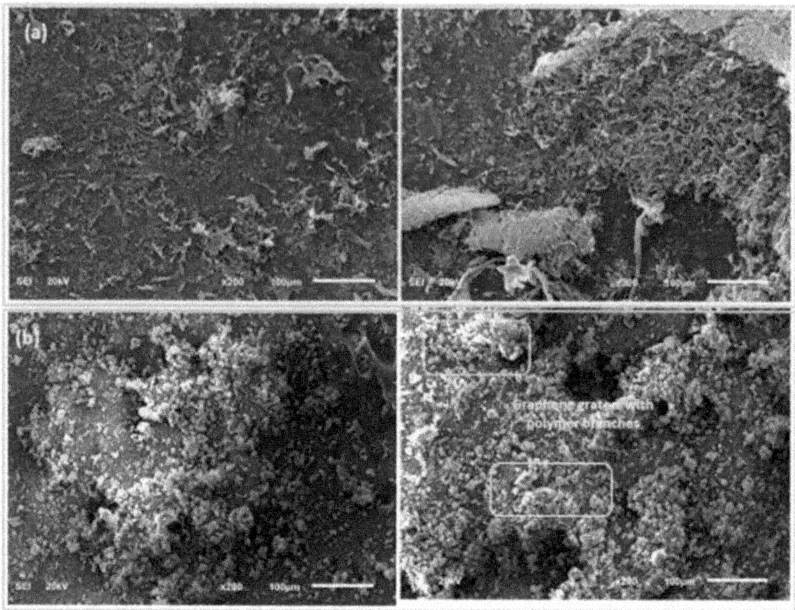

FIGURE 2.18
SEM images of (a) graphene oxide (GO) and (b) PEI/GO.

Source: Al-Yaari & Saleh, 2023, with permission.

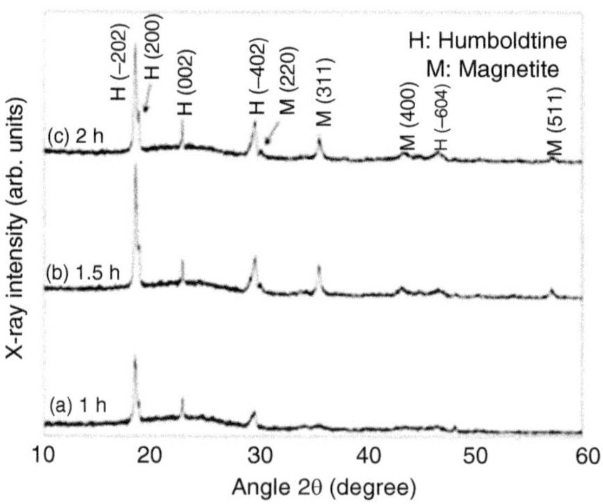

FIGURE 2.19
XRD of magnetic bamboo charcoal–iron oxide (BC/Fe) nanocomposite adsorbents synthesized via solvothermal process at different reaction times

Source: Nishioka & Sen, 2019; Sen, 2023b, with permission.

Biswas et al. (2020c) explored MB dye removal using a CASB composite adsorbent in a semi-fluidized bed adsorption column. XRD analysis (Figure 2.20) revealed that CASB had a crystallinity index (CI) of 21.62%, indicating a largely amorphous nature. The XRD pattern of the CASB adsorbent showed peak intensities that corresponded to the crystal structure of sugarcane bagasse biochar.

Additionally, Biswas et al. (2019a) synthesized a hybrid calcium alginate–pinecone biochar (CAPC) adsorbent and tested its capacity for Cu^{2+} and Ni^{2+} adsorption. XRD analysis (Figure 2.21) showed an amorphous to

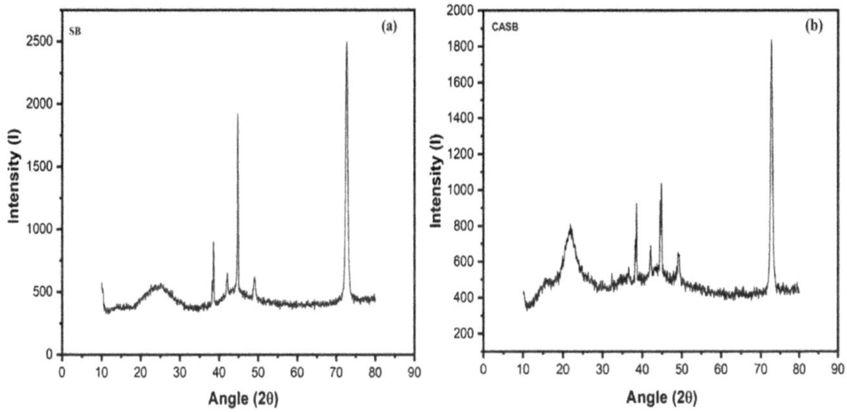

FIGURE 2.20
XRD patterns of (a) sugarcane bagasse biochar (SB) and (b) CASB composite

Source: Biswas et al., 2020c, with permission.

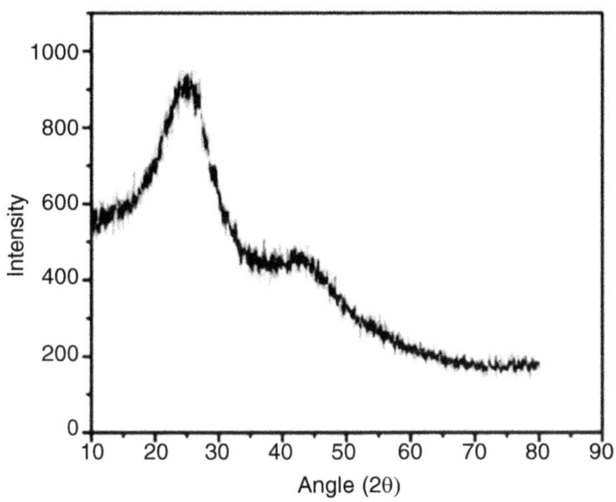

FIGURE 2.21
XRD pattern of calcium alginate–pinecone biochar (CAPC) adsorbent

Source: Biswas et al., 2019a, with permission.

semicrystalline structure, with peaks at 2θ values of 18°, 24°, 38°, and 45°, indicating the presence of hydrated crystals, which enhanced the adsorption capacity of the adsorbent (Biswas et al., 2019a).

Further XRD analysis of EB biochar, pinecone biochar (PB), and kaolin clay was conducted by Dawood et al. (2017b). Figure 2.22 shows that EB biochar exhibits peaks indicating organic matter and crystalline cellulose, PB biochar displays a silica peak (quartz form), and kaolin clay presents peaks corresponding to silica and titanium oxides, supporting its adsorptive capacity for pollutants (Dawood, 2018).

Ghosh et al. (2023) synthesized magnetic iron oxide nanoparticles modified with *Justicia adhatoda* leaf extract (JA-MIONs) for the removal of textile dyes and antibiotics from aqueous solutions. XRD analysis, shown in Figure 2.23, confirms the presence of Fe_3O_4 through distinctive crystalline phases and narrow peaks, indicating the effectiveness of JA-MIONs in adsorption.

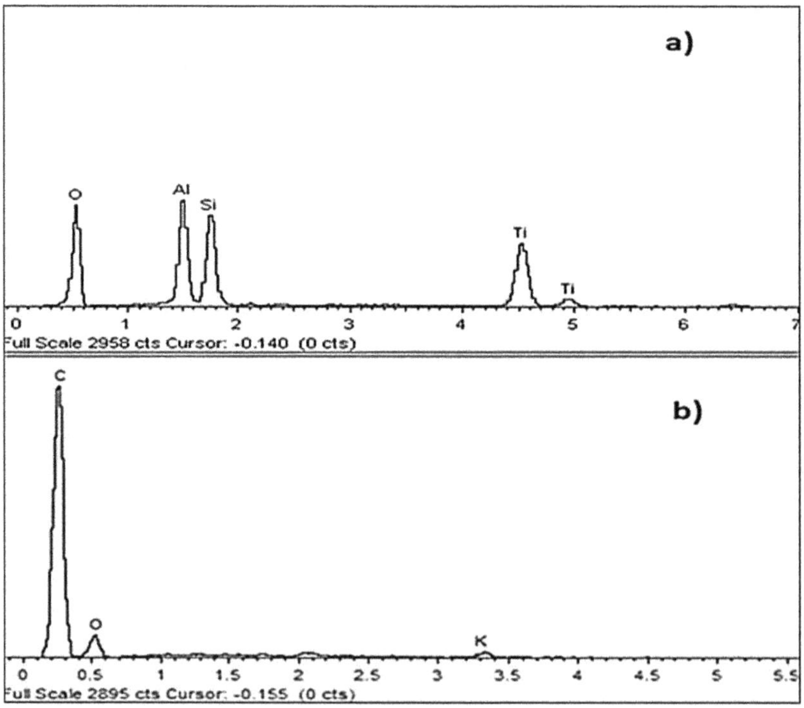

FIGURE 2.22
XRD analysis of (a) EB biochar, (b) pinecone biochar

Source: Dawood et al., 2017b, with permission.

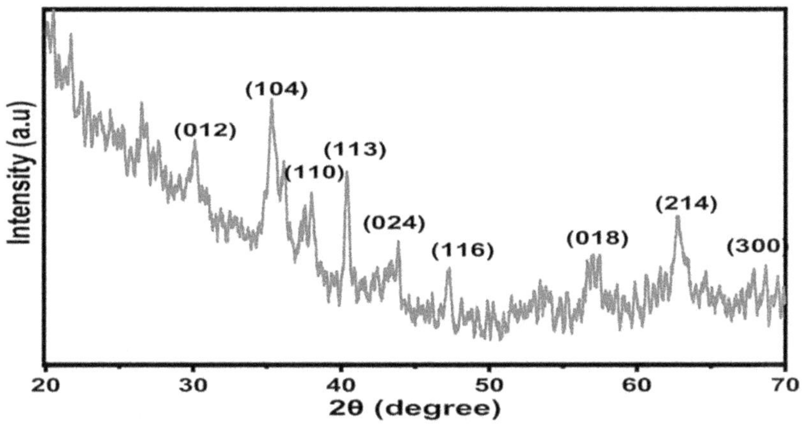

FIGURE 2.23
XRD pattern of JA-MIONs adsorbent

Source: Ghosh et al., 2023, with permission.

2.2.5.2.3 FTIR (Fourier Transform Infrared) Spectroscopy Analysis

FTIR analysis of adsorbent materials reveals surface functional groups essential for adsorbate adsorption, often forming adsorbent-adsorbate surface complexes. Functional groups like hydroxyl, carbonyl, phenolic, and carboxyl in biomass adsorbents facilitate the selective adsorption of gaseous and water pollutants (Afroza & Sen, 2018). Figure 2.24(a) shows the FTIR spectrum for pinecone biochar and its composite adsorbent, CAPC (calcium alginate pinecone biochar) (Biswas et al., 2019c, 2019c, 2020a). The FTIR spectrum of sugarcane bagasse biochar and its composite, CASB (calcium alginate sugarcane bagasse biochar), is presented in Figure 2.24(b) (Biswas et al., 2020c). Key findings include an O–H stretching peak at 3650–3748 cm^{-1}, present in pinecone biochar but absent in CAPC, and peaks at 2850–2924 cm^{-1} for C–H and N–H stretching. A weak peak around 2300 cm^{-1} indicates O=C=O groups, while peaks in the range 1000–1744 cm^{-1} correspond to C=C, N–H, and C–O stretching. In Figure 2.24(b), shifts and new peaks signify modifications in functional groups, indicating the presence of N–H stretching at 3446 cm^{-1} and C–H and N–H groups at 2854–2925 cm^{-1}, further supporting adsorption capacity.

Figure 2.25 presents the FTIR spectra for EB biochar before and after MB dye adsorption (Dawood, 2018; Dawood and Sen, 2014). Before adsorption, peaks at ~3337 cm^{-1} indicate O–H vibrations associated with alcohols, phenols, and carboxylic acids. Peaks at 1607 cm^{-1} and 1315 cm^{-1} correspond to N–H bending, with a small peak at 2162 cm^{-1} indicating C≡C stretching. After MB dye adsorption, shifts in these peaks—1607 to 1591 cm^{-1} and 1315 to 1325 cm^{-1}—suggest functional group interactions with MB molecules.

The FTIR analysis of kaolin clay before and after MB dye adsorption is shown in Figure 2.26. Sharp peaks at 3687 and 3619 cm^{-1} signify O–H and

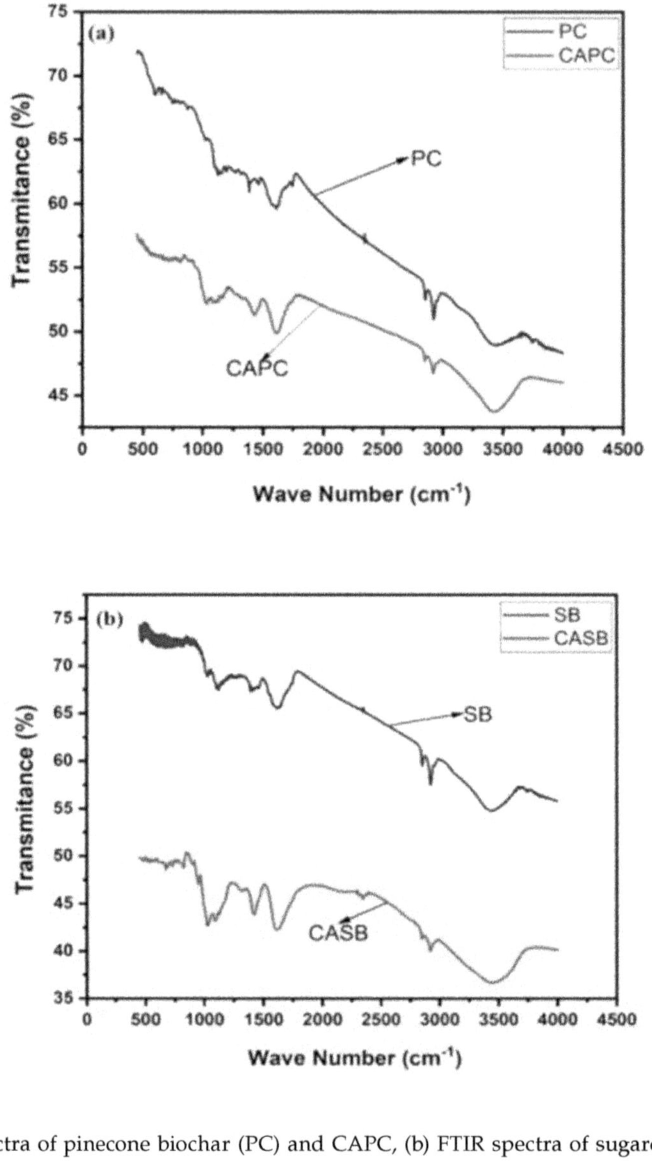

FIGURE 2.24
(a) FTIR spectra of pinecone biochar (PC) and CAPC, (b) FTIR spectra of sugarcane bagasse biochar (SB) and CASB

Source: Biswas et al., 2019a, 2019c, 2020c, with permission.

Si–O stretching vibrations, while peaks at 1023 cm^{-1}, 994 cm^{-1}, and 909 cm^{-1} indicate Si–O stretching and O–H deformation (Mukherjee et al., 2015). New weak peaks at 1605 cm^{-1} and 1391 cm^{-1}, observed post-adsorption, correspond to C=N and multiple ring stretching, confirming MB dye accumulation on kaolin surfaces (Dawood, 2017a, 2018).

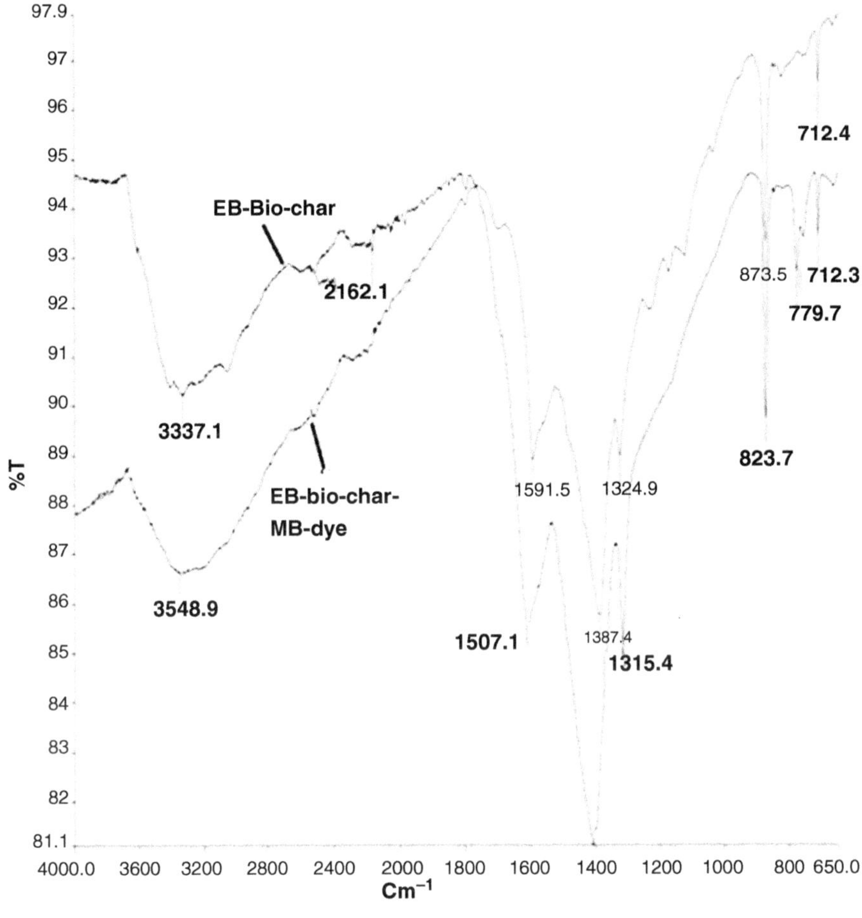

FIGURE 2.25
FTIR spectra of EB biochar before and after MB dye adsorption

Source: Dawood, 2018; Dawood and Sen, 2014, with permission.

Figure 2.27 displays FTIR spectra for raw EB and NaOH-modified EB powder (Afroze, 2016; Afroza et al., 2016a, 2016b). Key bands at 3314.9 cm^{-1} indicate –OH groups, while peaks at 1617.4 cm^{-1} and 1317.8 cm^{-1} correspond to C=O and N=O groups, respectively. For NaOH-modified EB, the slight shifts and intensity changes in peaks reflect increased surface area and pore structure modifications due to NaOH treatment, enhancing adsorption potential (Afroze, 2016).

2.2.5.2.4 Elemental Analysis of Select Adsorbents

CHNS-O analysis is instrumental in determining the carbon, hydrogen, nitrogen, sulphur, and oxygen content of adsorbents, providing insights

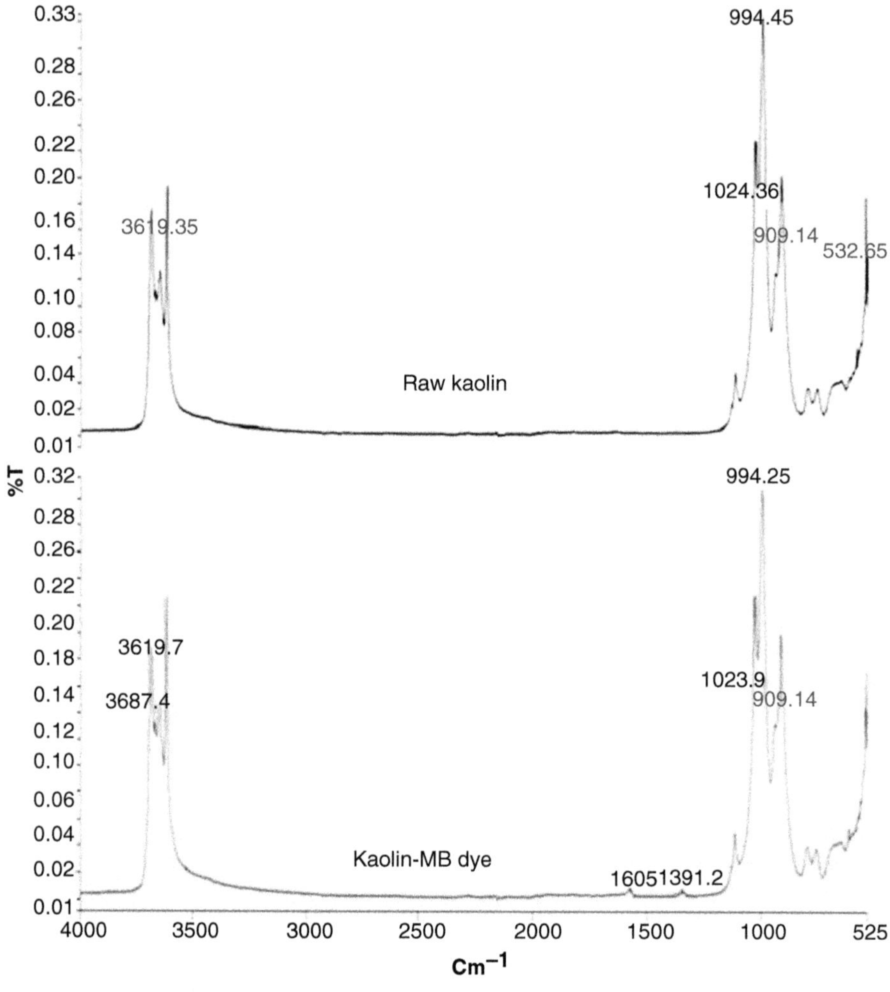

FIGURE 2.26
FTIR spectra of kaolin clay before and after MB dye adsorption

Source: Dawood et al., 2017a; Dawood, 2018 with permission.

into chemical compositions that influence adsorption behaviour (Afroze & Sen, 2018). Agricultural solid waste-based adsorbents, primarily composed of lignin, cellulose, and hemicellulose, exhibit diverse functional groups facilitating the binding process during adsorption (Afroze & Sen, 2018; Sen, 2015).

Table 2.1 outlines the chemical and mineralogical composition of various clay minerals, such as kaolinite, illite, montmorillonite, and chlorite, which are integral to adsorption processes (Khilar & Fogler, 1998).

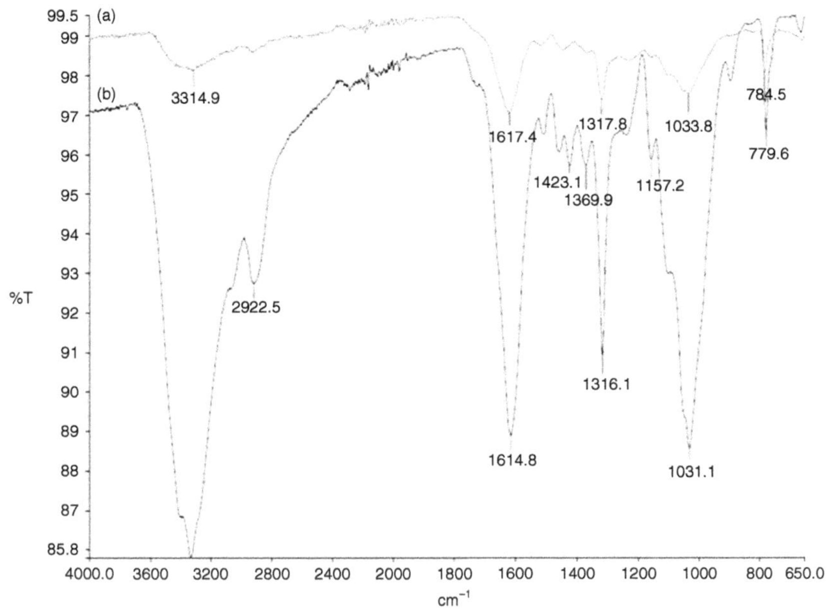

FIGURE 2.27
FTIR spectra of (a) raw and (b) NaOH-modified eucalyptus bark powder

Source: Afroza et al., 2016b, with permission.

TABLE 2.1
Chemical composition range of clay minerals (Courtesy: Khilar & Fogler, 1998)

Clay	SiO_2 (%)	Al_2O_3 (%)	Fe_2O_3 (%)	TiO_2 (%)	CaO (%)	MgO (%)	K_2O (%)	Na_2O (%)
Kaolin	40–49	35–40	0–13	0–2	0–0.8	0–0.5	0–0.10	0–0.2
Illite	50–56	18–31	2–5	0–0.8	0–0.2	1–4	-	0–1
Chlorite	31–38	18–20	-	-	-	35–38	-	-
Montmorillonite	45–55	19–50	0–3	-	-	0–2	-	-

The CHNS-O analyser results for elemental composition in various biochar, charcoal, and activated carbon adsorbents are summarized in Table 2.2 (Afroze & Sen, 2018; Afroze, 2016; Dawood, 2018).

The composition of lignocellulosic materials, which influence adsorptive properties, varies significantly across different agricultural wastes. Table 2.3 summarizes the cellulose, hemicellulose, and lignin content in several biomass adsorbents (Afroze & Sen, 2018).

2.2.5.2.5 *Surface Charge Density at the Solid/Liquid Interface (Point of Zero Charge, pH_{pzc})*

Understanding surface charge and surface potential of adsorbent particles in aqueous solutions is crucial for adsorption, as the uptake of ionic species is

TABLE 2.2

Elemental compositions (CHNS-O) of raw and agricultural waste-based adsorbents

Adsorbent	C (%)	H (%)	N (%)	O (%)	S (%)	Reference
Pinecone biochar (PC)	73.27	2.92	1.00	48.11	0	Biswas (2020)
Calcium alginate-PC (CAPC)	61.87	2.46	1.41	44.26	0	Biswas (2020)
Sugar bagasse biochar (SB)	84.43	2.28	0.93	35.06	0	Biswas (2020)
Calcium alginate-SB (CASB)	77.27	2.82	1.16	22.63	0	Biswas (2020)
Raw eucalyptus bark (EB)	42.32	5.69	0.24	-	0	Afroze (2016)
Eucalyptus EB-biochar	53.40	2.10	0.27	-	-	Dawood (2018)
Bamboo biochar	63.60	3.70	0.60	-	-	Li et al. (2016)
Banana biochar	71.40	6.30	0.60	-	-	Zhou et al. (2017)
Raw pinecone	47.10	5.50	0.36	-	-	Dawood (2018)
Commercial charcoal	73.60	4.80	1.40	-	0	Isahak et al. (2012)
Rice straw biochar	56.90	0.90	0.00	-	0	Sewu et al. (2017b)
Palm shell activated carbon	60.00	0.30	0.03	-	-	Yin et al. (2007)

TABLE 2.3

Chemical composition of selected biomass adsorbents

Adsorbent	Cellulose (%)	Hemicellulose (%)	Lignin (%)
Leaves	15–25	70–80	5–10
Wheat straw	30–35	26–32	16–21
Wood bark	25.2	30.3	44.5
Sugarcane bagasse	32–44	25–35	19–24
Pine	48.1	23.5	28.4
Barley straw	30–35	24–29	14–15
Cotton waste	80–95	5–20	-
Bamboo	26–43	15–26	21–31
Grass	30–40	35–50	10–25
Corncob	35–45	35–45	5–15
Olive stone	30–35	20–30	20–25
Nutshell and stone	25–35	25–30	30–40
Eucalyptus	52.7	31.9	15.4

influenced by the adsorbent's electrical properties at the solid/liquid interface. Surfaces of oxide adsorbent particles or silicate minerals in contact with aqueous solutions can become charged by the adsorption or desorption of H^+ ions (Sen, 2001). An electrical double layer forms at the solid/liquid interface through the adsorption of counter ions from the solution.

The **site dissociation model**, proposed by Healy and White, characterizes surface charge using the following reactions (Cornell & Schwartzman, 1998; Sen, 2001):

$$MOH_2^+(\text{surf}) = .MOH(\text{surf}) + H^+(\text{aq}) \tag{2.1}$$

and

$$MOH(\text{surf}) = MO(\text{surf}) + H^+(\text{aq}) \tag{2.2}$$

In this system H^+ and OH^- are potential determining ions and the surface charge and surface potential are defined by the pH of the solution. There is a particular pH at which the surface charge becomes zero, i.e. the point of zero charge (pH_{pzc}) or the isoelectric point (iep). At this point the extent of adsorption of positively charged species equals that of negatively charged species. It can be measured by electrophoresis as the pH at which there is no motion of the particles in the electric field (Cornel & Schwertmann, 1998; Sen, 2001). In general iron oxides is a strong adsorbent and its pH_{pze}'s in the pH range of 6–10 (Cornell & Schwertmann, 1998). Adsorption of cations is favoured at $pH > pH_{pzc}$, while adsorption of anions is favoured at $pH < pH_{pzc}$ (Afroze, 2016; Dawood, 2018).

The more relevant and useful method of surface charge characterization of adsorbent particles is, however, the measurement of **zeta potential of the particle**. Hussain et al. (1996) measured zeta potential of clay minerals such as kaolinite, illite, and chlorite clay minerals against pH and are found to have negative zeta potential in the pH range of 2.5 to 11. The kaolinite is the most negative clay, ranging in zeta potential value from −24 to −49.5 mV. Parks (1967) summarized the factors controlling the sign and magnitude of surface charge of oxide and mineral oxides, especially hydrous metal oxides which are very much used in the adsorptive removal of inorganics and organics from aqueous phase as strong adsorbent. According to Parks (1967), in acid solutions, the surface charge is positive and therefore the hydrous metal oxides act as anion exchangers. In basic solutions the surface charge is negative, and they are cation exchange particles. In neutral solutions, the surface charge is mixed, and the particles show a limited capacity for exchange of both cations and anions.

The pH_{pzc} for the various adsorbents can be determined by solid addition method described by Mall, Srivastava et al. (2006). This technique is based on estimating the concentrations of H^+ and OH^- ions, which determine the electric potential develop. As per this solid addition method,

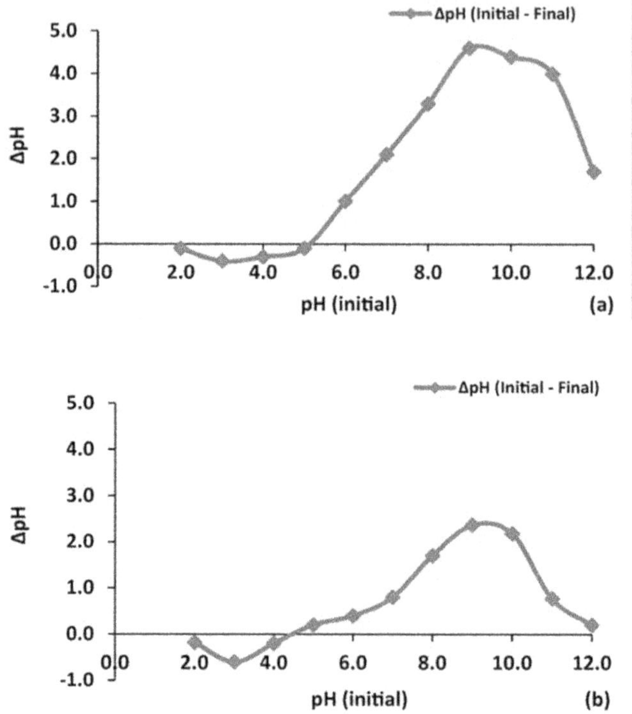

FIGURE 2.28
Point of zero charge (pH_{pzc}) of (a) raw eucalyptus bark and (b) NaOH-modified eucalyptus bark biomass

Source: taken from Afroze, 2016; Afroze & Sen, 2018 with written permission.

Afroze (2016) was measured the pH_{pzc} of raw EB and base NaOH-modified eucalyptus bar adsorbents, which is presented in Figure 2.28a–b. In this method, a series of 0.1 M, KNO_3 solutions (50 mL each) were prepared, and their initial pH was adjusted at 2–12 by addition of 0.1 M HCl or NaOH solution. Then, 0.1 g of adsorbent was added to each prepared solution and the suspensions were shaken and the solution was kept for a period of 48 hrs. The final pH of each solution was recorded with a pH metre, and the difference between initial and final pH (ΔpH) (Y-axis) was plotted against initial pH values (X-axis) (Figure 2.28). The point of intersection of this curve yielded pH_{pzc} (Afroze, 2016). From Figure 2.28a, pH_{pzc} of raw EB biomass was found to be around 5, whereas base modified the same EB biomass showed a pH_{pzc} value of 4.5 in Figure 2.28b (Afroze, 2016). Recently, Sen (2023a) reviewed and compiled the pH_{pzc} values of various agricultural solid waste-based adsorbents used in the removal of heavy metals ions from water.

2.2.5.3 Physical Properties of Clay Minerals and Agricultural Solid Waste-Based Adsorbents in Adsorption: Particle Size, BET Surface Area Analysis, and Bulk Density

In adsorption processes, particle size, shape, and surface area are critical factors. A larger surface area, typically from smaller particle sizes, improves adsorption efficiency, whereas larger particles with lower surface areas are less effective.

The particle size distribution of raw EB biomass, analysed by the Malvern Master Sizer, is presented in Figure 2.29. Results showed a surface-weighted mean of 25.26 μm, a volume-weighted mean of 79.67 μm, and a specific surface area of 0.238 m^2/g for raw EB powder used in aqueous phase adsorption of inorganics and organics (Afroze, 2016).

Further analyses by Dawood et al. 2017b, on EB biochar, pinecone biochar, and kaolin clay, are shown in Figure 2.30. The EB biochar demonstrated a specific surface area of 0.56 m^2/g with a particle size of 10.68 μm, while pinecone biochar had 0.26 m^2/g and 23.0 μm, respectively. Kaolin showed a specific surface area of 1.80 m^2/g and a particle size of 6.38 μm (Dawood, 2018).

BET surface area analysis is critical as larger surface areas increase adsorbate-adsorbent interactions, enhancing adsorption efficiency. Table 2.4 details the BET surface areas for various adsorbents, including biochar, activated carbon, and clay minerals, as reported by different researchers.

Bulk density is crucial for adsorbent selection in continuous adsorption columns, impacting the column's performance. A bulk density exceeding 0.25 g/cm^3 is generally preferred in field applications (Biswas, 2020; Dawood, 2018). Bulk density is calculated using the formula:

$$\text{Bulk Density} = \text{mass of dry sample (g)} / \text{ total volume used (mL)} \quad (2.3)$$

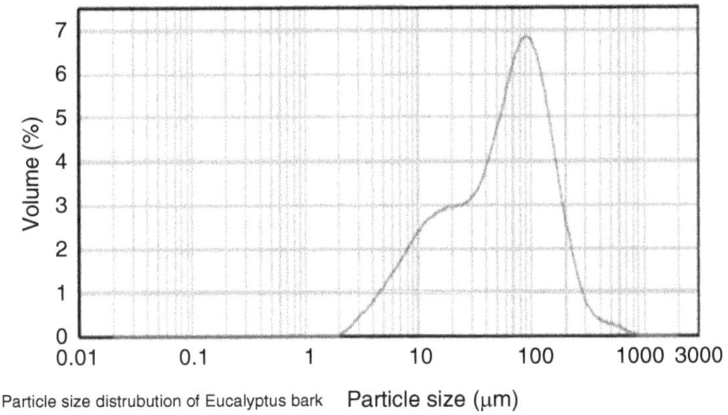

FIGURE 2.29
Particle size distribution of raw eucalyptus bark (EB)

Source: Afroze, 2016.

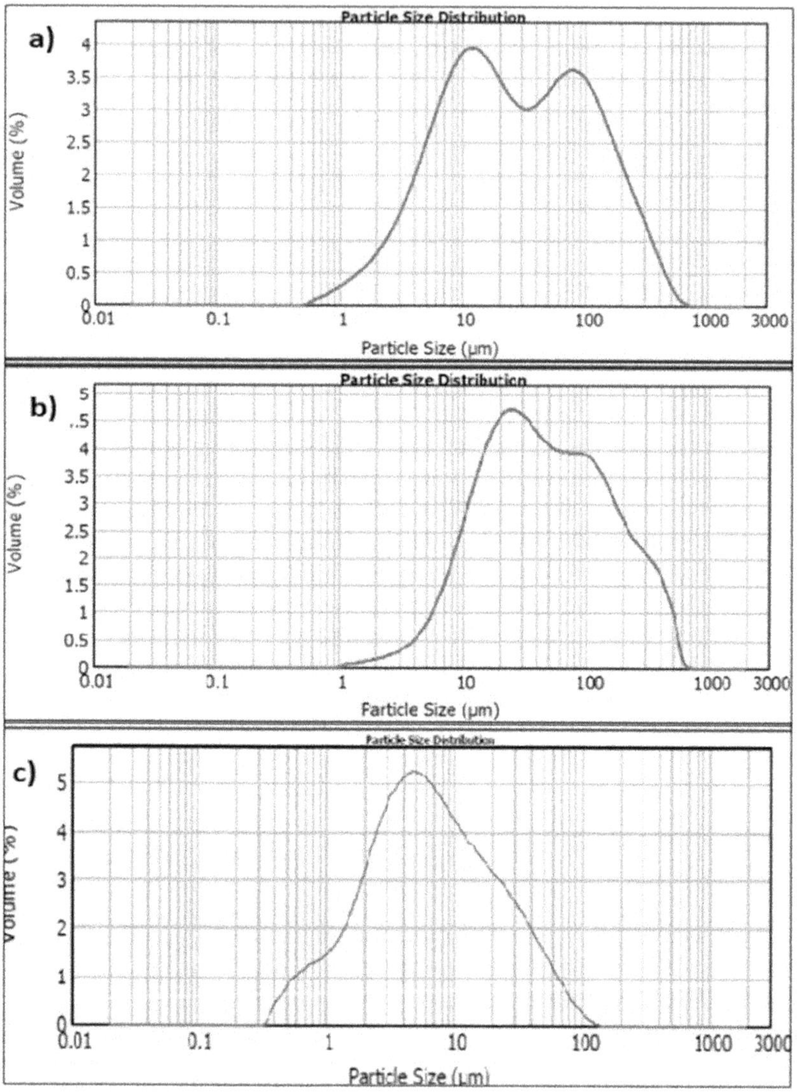

FIGURE 2.30
Malvern particle sizer analysis for (a) EB biochar, (b) pine-cone biochar, and (c) kaolin clay

Source: Dawood, 2018; Dawood et al. 2017b with written permition.

Table 2.5 presented the bulk density of few agricultural solid wastes-based adsorbent used in the removal of water pollutants.

TABLE 2.4

BET surface area of few common adsorbents used in the adsorptive removal of water and air pollutants reported by various investigators

Adsorbent	BET surface area (m^2/g)	Reference
Pinecone biochar (PC)	144.94	Biswas et al. (2020a)
Raw pinecone	15.5	Dawood (2018)
Calcium alginate-pinecone biochar	163.28	Biswas (2020)
Sugarcane bagasse biochar (SB)	391.42	Biswas (2020)
Calcium alginate-SB biochar (CASB)	200.14	Biswas (2020)
Raw eucalyptus bark (EB)	6.55	Afroze (2016)
Base-treated eucalyptus bark	20.13	Afroze (2016)
EB-biochar	73.0	Dawood (2018)
Kaolin clay	30.5	Dawood (2018)
Banana peel biochar	31.7	Zhou et al. (2017)
Chitosan activated carbon	318.4	Marrakchi et al. (2017)
Fly ash (FA)	200.0	Ge et al. (2018)
Activated carbon (AC)	1202.0	Mohamed et al. (2024)
TESPA/carbon nanotube (CNT)/AC	442.0	Mohamed et al. (2024)
Activated fly ash carbon (AC)-untreated	818	Kaithwas et al. (2012)
Activated fly ash carbon with DEA	265	Kaithwas et al. (2012)

TABLE 2.5

Bulk density of some agricultural solid wastes-based adsorbents and clay minerals used in the adsorptive removal of water pollutants reported by various investigators

Adsorbent	Bulk density (g/cm^3)	Reference
Pinecone biochar (PC)	0.31	Biswas et al. (2019b, 2019c)
Calcium alginate-pine cone biochar	0.41	Biswas et al. (2019b, 2019c)
Sugarcane bagasse biochar (SB)	0.29	Biswas et al. (2019b, 2019c)
Calcium alginate-sugarcane bagasse (SB)	0.39	Biswas et al. (2019b, 2019c)
Raw eucalyptus bark (EB)	0.39	Afroze (2016)
NaOH-treated eucalyptus bark	0.27	Afroze (2016)
Eucalyptus bark (EB)-biochar	0.42	Dawood (2018)
Pinecone activated carbon	0.36	Dawood et al. (2014)
Kaolin clay minerals	0.67	Dawood (2018)

2.3 State-of-the-Art on the Application of Various Adsorbents in the Removal of Water and Gaseous Pollutants

CAC adsorbents are the most used commercial adsorbents for removing gaseous and water pollutants due to their high specific surface area, porous structure, and functional groups, such as carboxyl, carbonyl, phenol, and amine, which bind adsorbate molecules to form surface complexes at the gas/solid or liquid/solid interface (Deliyanni et al., 2015; Fu & Wang, 2011). However, current research focuses on developing solid waste-based adsorbents as alternatives to costly commercial CAC adsorbents, making the adsorptive separation process a more effective, popular, and sustainable option for pollutant remediation (Sen, 2023a). Based on precursor materials, adsorbents are classified into agricultural by-product solid waste-based adsorbents, aquatic and terrestrial biomass-based adsorbents, zeolites, clay materials, industrial solid waste-based adsorbents, and CNTs (Burakov et al., 2018).

2.3.1 Agricultural By-product Solid Waste-Based Adsorbents

Raw or modified agricultural by-product solid wastes, such as leaves, bark, fibres, seeds, fruit peels, and sawdust, have been successfully used as potential adsorbents for removing various water and gaseous pollutants (Figures 2.4–2.6). Literature demonstrates that modifying raw agricultural adsorbents improves the surface properties of agricultural wastes, which enhances metal ion sequestration and dye removal from wastewater (Afroze & Sen, 2018; Gautam, Mudhoo et al., 2014). Rice bran and rice straw have been used for removing Cu^{2+} from aqueous solutions, with maximum adsorption capacities of 21 and 18.4 mg/g, respectively (Singha & Das, 2013). Complete removal of Cr^{6+} was achieved using rice husk waste adsorbent at low pH (<3) with an adsorption capacity of 8.5 mg/g. Similarly, metal ions Pb^{2+} and Cr^{6+} were effectively removed from water by peanut shell residue adsorbent (Ahmad et al., 2017; Sen, 2023a). Several investigations report the ability of cashew nutshell adsorbents in removing heavy metal ions from water. For example, 20 mg/g of Cu^{2+} was removed from the aqueous phase using this cashew nutshell adsorbent (SenthilKumar et al., 2011). In another study, Ni^{2+} removal reached 75%, with a maximum capacity of 18.9 mg/g (Kumar et al., 2011). Groundnut shells were successfully applied for the removal of Cu^{2+}, Ni^{2+}, and Zn^{2+}, with adsorption capacities of 4.9, 8.05, and 11 mg/g, respectively (Shukla & Pai, 2005).

Fruit wastes, such as lemon peel, orange peel, and banana peel, have also been used as effective adsorbents for removing inorganics/organics from aqueous phases (Thirumavalavan et al., 2010). Anastopoulos et al. (2017) reviewed and compiled various coffee-based adsorbents, including coffee grounds, coffee residues, spent coffee grains, and coffee husks, in the removal of aqueous phase heavy metal ions under different experimental conditions. Sen (2023a) reviewed and compiled many agricultural solid waste-based adsorbents successfully used in the removal of aqueous phase heavy metal ions by adsorption. Table 2.6 presents a compilation of various raw and modified agricultural by-product solid waste-based adsorbents used for removing organics (dyes, phenol, antibiotics) along with their maximum adsorption capacities from isotherm studies (Biswas, 2020; Dawood, 2018; Sen, 2023a). Table 2.7 presents the results for various raw and modified agricultural solid waste-based adsorbents used in the removal of

TABLE 2.6

Adsorption capacities q_m (mg/g) of various raw and modified agricultural solid waste-based adsorbent materials for organic dyes and phenolic compounds adsorption (Biswas, 2020; Dawood, 2018)

Adsorbent	Adsorbate	Max adsorption capacity, q_m (mg/g)	References
Bamboo-based AC	Methylene blue	576	Yao et al. (2024)
Forest waste-based AC	Methylene blue	198.01	Yildiz (2024)
Mango leaf	Methylene Blue	156	Uddin et al. (2017)
Durian peel	Acid Blue 25	26.6	Dahri et al. (2016)
Peanut husk	Indosol orange	79.5	Sadaf et al. (2014)
Peanut hulls	Golden yellow	35.7	Nawaz et al. (2014)
Bamboo leaves	Methylene blue	54.2	Zhu et al. (2016)
Mango seed	Methylene blue	25.4	Senthil Kumar et al. (2014)
Coffee waste	Toluidine blue	142.5	Lafi et al. (2014)
Pinecone	Acid black 26	62.9	Mahmoodi et al. (2011)
Pine tree leaves	Basic red 46	79.1	Deniz and Karaman (2011)
Palm kernel fibre	Crystal violet	78.9	El-Sayed and Owes (2011)
Carbon nanofibres	2,4 dichlorophenol	483.09	Zhang et al. (2018)
Calcium-alginate-PC	Phenol	16.33	Biswas (2020)
Calcium-alginate-SB	Phenol	20.20	Biswas et al. (2020b)
Calcium-alginate-SB	Methylene blue	68.4	Biswas et al. (2020c)
Calcium-alginate-PC	Methylene blue	70.66	Biswas (2020)
Raw pinecone	Congo red	19.18	Dawood and Sen (2014)
Acid-treated pinecone	Congo red	40.19	Dawood and Sen (2014)
Pinecone-based AC	Congo red	76.45	Dawood and Sen (2014)
Bamboo-char-iron oxide	Methylene blue	111.11	Sen (2023b)
Pinecone	Methylene blue	109.89	Sen et al. (2011)
Palm shell	Reactive red 141	14.0	Dawood and Sen (2014)
Pine tree leaves	Methylene blue	126.6	Dawood et al. (2014)
Raw coffee residue	Basic blue 3G	251.0	Dawood et al. (2014)
Raw peanut hull	Reactive black 5	55.5	Afroze and Sen (2018)

(Continued)

TABLE 2.6

(Continued)

Adsorbent	Adsorbate	Max adsorption capacity, q_m (mg/g)	References
Chemically treated cucumber peel	Direct blue 71	129.87	Saygili and Guzel (2017)
Chemically treated durian leaf	Methylene blue	125.0	Hussin et al. (2015)
Chemically treated neem bark	Malachite green	500.00	Sharma et al. (2014)
Iron oxide nanoparticles-leaf extract	Antibiotic TCH	200.0	Ghosh et al. (2023)
Pecan nutshell biochar	Reactive red 141	130	Zazycki et al. (2018)
Rice straw biochar	Crystal violet	620.3	Sewu et al. (2017b)
Bamboo-hydro-char	Congo red	33.7	Li et al. (2016)
Java plum leaves	Malachite green (MG)	140	Gul et al. (2024)
Date seed-based AC	Bismark brown R	192.31	Sivalingam and Sowmiya (2024)

TABLE 2.7

Adsorption capacities q_m (mg/g) of recently reported raw and modified agricultural solid waste-based adsorbent materials for inorganics (mainly heavy metal ions) adsorption (Biswas, 2020; Dawood, 2018; Sen, 2023a)

Adsorbent	Adsorbate	Max adsorption capacity, q_m (mg/g)	References
Acid-modified	Mn^{2+}	45.12	Sivalingam and Gopal (2024)
Aloe vera plant	Fe^{3+}	41.28	Sivalingam and Gopal (2024)
Date seed-based AC	Zn^{2+}	15.55	Sivalingam and Sowmiya (2024)
Modified orange peels	Cu^{2+}	289.0	Chakraborty et al. (2022)
Jackfruit peels	Cu^{2+}	17.5	Ibrahim et al. (2020); Ayob et al. (2021)
	Pb^{2+}	10.0	
	Cd^{2+}	20.0	
	Mn^{2+}	76.9	
Coconut husk	Zn^{2+}	338.0	Malik and Dahiya (2017)
Modified eucalyptus bark	Zn^{2+}	250	Afroze et al. (2016b)
Avocado seed	Cr^{6+}	35.5	Ahmet and Danish (2022)

(Continued)

TABLE 2.7
(Continued)

Adsorbent	Adsorbate	Max adsorption capacity, q_m (mg/g)	References
Date palm empty fruit bunch	Cr^{6+}	70.49	Rambabu (2020); Sen (2023a)
Raw coconut shell	Cd^{2+}	3.5	Paranavithana et al. (2016)
Raw coconut shell	Pb^{2+}	13.4	Paranavithana et al. (2016)
Coffee pulp	Cr^{6+}	13.48	Ayob et al. (2016)
Modified coconut husk	Ni^{2+}	404.3	Malik and Dahiya (2017)
Modified coconut husk	Cu^{2+}	443.2	Malik and Dahiya (2017)
Sugarcane bagasse	Mn^{2+}	0.43	Anastopoulos et al. (2017)
Coconut tree sawdust	Pb^{2+}	25	Putra et al. (2014)
	Zn^{2+}	23.8	
Modified rice husk	Hg^{2+}	89	Song et al. (2014)
Treated rice straw	Cr^{6+}	15.8	Wu et al. (2016)
Magnetic bamboo charcoal	As^{3+}	19.6	Nomura et al. (2017)
Sugarcane bagasse	Hg^{2+}	97.58	Chakraborty et al. (2020)

inorganic heavy metal ions from water, along with their maximum adsorption capacity, q_m(mg/g).

Nowicki et al. (2016) synthesized micro/mesoporous activated carbon from biodegradable waste and successfully used it in the adsorptive separation of gaseous pollutants nitrogen dioxide (NO_2) and hydrogen sulphide (H_2S), achieving adsorption capacities of 83 mg NO_2/g and 215 mg H_2S/g. Wei et al. (2016) reported that waste ion exchange resin-based chemically activated microporous activated carbon was able to remove pure carbon dioxide (CO_2) by adsorption through TGA testing at 30°C under environmental pressure, with a maximum adsorption capacity of 81.24 mg CO_2/g. Marigold-based activated carbonaceous adsorbents achieved a maximum adsorption capacity of 102.1 mg NO_2/g under specified conditions (Bazan et al., 2016). Plum stone-based activated carbon was found to adsorb NO_2 at 65 mg/g (Nowicki et al., 2010), while sawdust-based activated carbon showed an adsorption capacity of 43.1 mg/g for NO_2 (Pietrzak, 2010).

2.3.2 Aquatic and Terrestrial Biomass-Based Bioadsorbent

Biological materials such as chitin, chitosan, peat, yeasts, fungi, and bacterial biomass are used as bioadsorbents to remove dyes and heavy metal ions from aqueous solutions (Afroze, 2016). Biosorbents are often more selective than traditional ion exchange resins and commercial activated carbons, capable of reducing dye concentrations to low levels (Ali, 2010; Rafatullah, Sulaiman et al., 2010). Numerous studies on the biosorption of dyes from aqueous solutions have been conducted globally, compiled by Asgher (Asgher & Bhatti, 2012). The bioadsorption of heavy metals from aqueous solutions is relatively new, with recent advancements reported by Ghosh et al. (Ghosh Dastidar et al., 2015). Diverse functional groups such as carboxyl, hydroxyl, sulphate, and amino groups present on the biosorbent are responsible for the binding mechanisms (Yagub, Sen et al., 2014). Various forms of trees, marine plants, and terrestrial biomass are inexpensive and readily available in large quantities. For instance, *Moringa Oleifera* (MO) has shown promising results in removing heavy metals like Cu^{2+}, Cd^{2+}, Cr^{3+}, Zn^{2+}, and Pb^{2+} from aqueous solutions (Kansal & Kumari, 2014; Shan et al., 2017). Tree fern is another plant-based adsorbent with widespread applications; an isotherm study showed it could remove Zn^{2+}, Cu^{2+}, and Pb^{2+} with adsorption capacities of 7.6, 10.6, and 39.8 mg/g, respectively (Ho et al., 2002).

Other tree-based waste adsorbents, like those from *Lagerstroemia speciosa*, an Indian native plant, have been investigated for Cr^{6+} removal, achieving a maximum adsorption capacity of 20.4 mg/g (Srivastava et al., 2015). Sawdust from various trees has also shown effective heavy metal removal. For example, maple sawdust removes over 80% of Cr^{6+} (Chen et al., 2010), and beech sawdust has demonstrated simultaneous removal of Cu^{2+}, Ni^{2+}, Cd^{2+}, and Zn^{2+} (Božić et al., 2013). Other tree-based adsorbents include poplar tree sawdust, teakwood sawdust, and Indian jujube tree (Ahmad et al., 2017).

Using algae in living and non-living forms is also effective for heavy metal ion uptake (Lamai et al., 2005). Non-living algae, due to its polymeric structure containing cellulose, glycoproteins, and similar compounds, are particularly effective (Volesky, 2007). Alginic acid or its salts from non-living algae are effective for metal ion adsorption, as they contain functional groups such as OH^- and COO^- that bind positively charged metal ions, forming surface complexes or enhancing electrostatic charge separation (Kaplan, 2013). Extensive studies have also examined algal biochar as a biosorbent with enhanced biosorption properties for removing toxic dyes and heavy metals (Davis et al., 2003).

2.3.3 Inorganic Adsorbent Materials

2.3.3.1 Clay Minerals

Among natural inorganic materials, clay minerals are well-known adsorbents due to their low cost, abundant availability, strong sorption properties, and ion exchange potential. Clays are effective in removing organic and inorganic contaminants due to their large surface area, high ion exchange capacity, and presence of Brønsted and Lewis acid sites, as well as their layered structure and chemical and mechanical stability (Aries & Sen, 2009). Clay minerals contain exchangeable ions, such as Ca^{2+}, Mg^{2+}, H^+, K^+, NH_4^+, Na^+, SO_4^{2-}, Cl^-, PO_4^{3-}, and NO_3^-, which can be exchanged with other cations

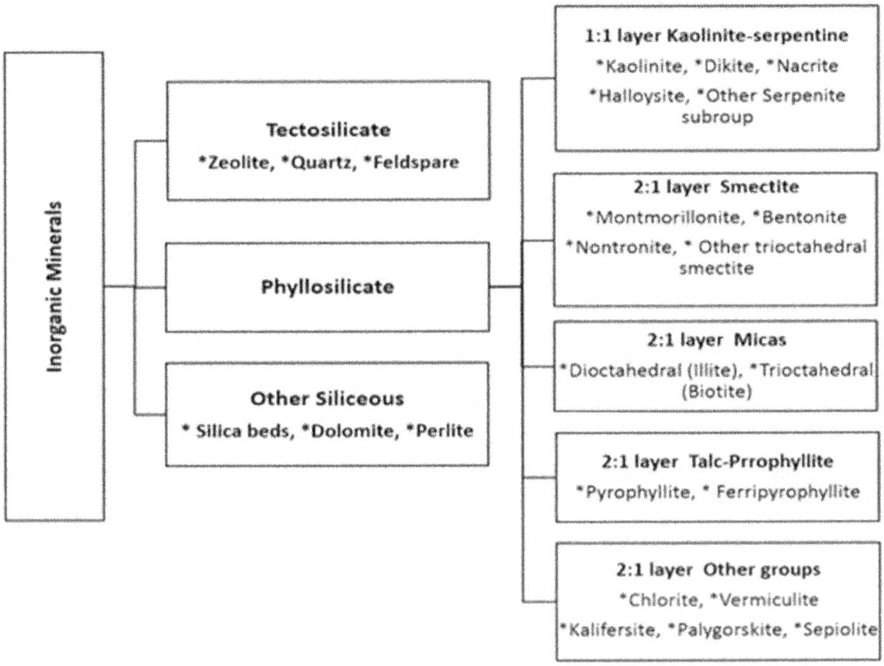

FIGURE 2.31
Various inorganic clay minerals

Source: Dawood, 2018; Brigatti et al., 2013, with written permission.

and anions without disrupting the clay's structure (Rafatullah, Sulaiman et al., 2010). Thus, clay minerals show a strong affinity for both cations and anions. Clays can also be modified to improve their sorption capacity through treatments, such as replacing cations or using heat and acid treatment (Gupta, 2009).

Clay minerals possess a layered structure and act as host materials for adsorbates and counter ions, playing a significant role in environmental contaminant removal by acting as natural scavengers. Inorganic clays and other silica materials can be categorized into various groups, including the kaolin-serpentine group (e.g., kaolinite, dickite, nacrite, halloysite), the talc and pyrophyllite group (e.g., pyrophyllite, ferripyrophyllite), the mica group (dioctahedral and trioctahedral forms), the smectite group (e.g., montmorillonite, bentonite), the chlorite group, 2:1 layer silicates (e.g., kalifersite, palygorskite, sepiolite), and the tectosilicate group, as shown in Figure 2.31 (Brigatti et al., 2013; Dawood, 2018; Yagub et al., 2014).

Table 2.8 demonstrates the removal capabilities of various clays for dyes and heavy metal ions from aqueous phases as reported by several researchers.

2.3.3.2 Zeolites

Zeolites are naturally occurring or industrially produced crystalline aluminosilicate materials. Zeolites are considered one of the best adsorbents due to their composition of interlinked alumina (AlO_4) and silica (SiO_4) structures (Choi et al., 2016). Their high surface area, hydrophilicity, and ion exchange capability make zeolites highly effective as adsorbents. Zeolites have a highly porous aluminosilicate framework with different cavity structures, consisting of a three-dimensional lattice with a negatively charged framework (Rafatullah, Sulaiman et al., 2010). This negative charge is balanced by exchangeable cations, which can be substituted with certain cations in solution. Zeolites have been extensively studied for their ability to remove trace contaminants such as heavy metals, phenols, and dyes (Afroze, 2016). Among over 40 natural varieties, clinoptilolite, a mineral from the heulandite group, is one of the most abundant and commonly studied (Yagub, Sen et al., 2014). Clinoptilolite is widely used to remove a range of heavy metals (Shaheen et al., 2012). Modification of natural zeolite may reduce particle size and increase surface area, enhancing adsorption efficiency. For instance, nano-sized NaX zeolite has been widely used to treat Cd^{2+}-laden wastewater (Ansari et al., 2014). In another study, Mg-modified zeolite achieved more than 98% removal efficiency for Ca^{2+} from aqueous solutions, attributed to its large surface area (Biswas, 2020). Biochar modified with sodium alginate and composite zeolitic imidazolate framework 8 (SA@ZIF-8/BC) was used for Pb^{2+} removal, with an adsorption capacity of 300 mg/g—2.36 times higher than that of unmodified biochar (Peng et al., 2024).

2.3.3.3 Siliceous Materials

Natural siliceous sorbents, such as silica beads, alunite, dolomite, and perlite, are increasingly used for wastewater treatment due to their low cost, abundance, and high sorption properties (Yagub, Sen et al., 2014). The porous structure, mechanical stability, chemical reactivity of the hydrophilic surface, and high surface area of silica beads make them ideal for pollutant removal applications (Crini, 2006). Consequently, siliceous materials are effective adsorbents for wastewater treatment, as demonstrated in several studies, summarized in Table 2.8.

2.3.3.4 Industrial Solid Waste

Industrial activities produce solid waste materials; some by-products are reused, while others are disposed of in landfills (Ruhma Rashid et al., 2021). Industrial solid wastes are generated at millions of tons worldwide daily. These materials are by-products with no economic value, making them very inexpensive. Improper disposal of these solid wastes poses a serious environmental threat and is thus a global concern (Kaithwas et al., 2012). Several industrial solid wastes exhibit promising adsorption capacities and are used

TABLE 2.8

Adsorption capacities, q_m (mg/g) of natural inorganic materials for dyes and heavy metal ions

Adsorbent	Dyes & metal ions	Max. Adsorption capacities, q_m (mg/g)	References
Illite-montmorillonite clay	Methylene blue	1.925×10^{-2} kg/kg	Mohamed et al. (2024)
Kaolinite	Crystal violet	20.64	Khan and Khan (2015)
Natural clay	Reactive red 120	30.3	Abidi, Errais et al. (2015)
Raw ball clay	Methylene blue	34.65	Auta and Hameed (2012)
Modified ball clay	Methylene blue	100	Auta and Hameed (2012)
Bentonite/zeolite composite	Methylene blue	36.2	Shaban et al. (2017)
	Congo red	46.3	Shaban et al. (2017)
Kaolin	Methylene blue	80.7	Dawood et al. (2017a)
Montmorillonite	Crystal violet	370.4	Sarma et al. (2016)
Acid-treated Montmorillonite	Crystal violet	384.6	Sarma et al. (2016)
Acid-modified zeolite	Methylene blue	2.1	Hor et al. (2016)
Meta-kaolin	Zn^{2+}	74.5	Kara et al. (2017)
	Ni^{2+}	46.2	Kara et al. (2017)
Bentonite	Al^{3+}	8.5	Chai et al. (2017)
Kaolin	Al^{3+}	1.2	Chai et al. (2017)
Sepiolite	Hg^{2+}	54.7	Vağzoğullar et al. (2015)
Modified sepiolite	Hg^{2+}	104.2	Vağzoğullar et al. (2015)
Kaolinite	Zn^{2+}	7.2	Joh and Mbadcam (2015)
Dolomite	Cu^{2+}	225.7	Qiu, Xie et al. (2015)

for removing toxic pollutants from water and wastewater. Industrial solid waste-based adsorbents, such as zeolite from fly ash (Lee & Jo, 2010), blast furnace slag, dust and sludge, and red mud from alumina industries, have been successfully applied in removing gaseous and water pollutants through adsorption (Bhatnagar et al., 2011; Kaithwas et al., 2012; Yadav et al., 2010). Fly ash, with its high alumina and silica content, is an inexpensive adsorbent suitable for adsorption (Bhatnagar & Sillanpää, 2010). Red mud, an aluminium industry effluent, has a porous, smooth, and dense surface texture, which makes it effective in removing various aquatic contaminants (Ruhma Rashid et al., 2021). Yadav et al. (2010) successfully used red mud for CO_2 sequestration. Solid residues from the steel industry have been tested as adsorbents for hydrogen sulphide at high concentrations, with notable results for thiocyanate removal from aqueous solutions (Das et al., 2007). The cardboard waste-activated carbon (CAC) yields a specific area of 196 m^2/g and provides a malachite green dye capacity of 134 mg/g, as reported by Alarboday et al.

(2024). Modification of coffee industry sludge-based adsorbent using acid treatment has also been effective for tartrazine dye removal from aqueous solution, achieving an adsorption capacity of 599.64 mg/g (Pesol et al., 2024).

2.3.3.5 Nano-Based Adsorbents

Recently, nanomaterials, nano-adsorbents, or nanoparticles have shown successful applications in several areas of interest, primarily in removing metallic pollutants from industrial wastewater (Rashid et al., 2021; Sharma et al., 2009). These materials are categorized into three main types: fullerenes, CNTs, and GOs. With the rise of nanotechnology, various carbon-based nanomaterials have been developed (Rashid et al., 2021; Smith & Rodrigues, 2015). Carbon and graphene-based nanomaterials, such as nanotubes, are available in functionalized and non-functionalized forms (Kuila et al., 2012). Fullerenes are closed-cage carbon structures with pentagonal or hexagonal configurations, and their adsorption characteristics make them effective adsorbents for purifying heavy metal-laden wastewater due to their high surface area and low tendency for agglomeration (Biswas, 2020; Scida et al., 2011). For example, fullerenes have been applied for Cu^{2+} removal with an adsorption capacity of 14.6 mg/g. Additionally, fullerene-doped activated carbon (AC) showed a 1.5- to 2.5-fold increase in adsorption capacity for Pb^{2+} and Cu^{2+} (Samonin et al., 2014).

Carbon nanotubes (CNTs), first introduced in 1991 (Iijima, 1991), have proven highly effective as adsorbents for pollutant removal from water, promoting a healthier environment (Yang et al., 2009). CNTs consist of single- or multi-layered cylinders formed from rolled graphite plates, with carbon atoms arranged in a hexagonal network. CNTs have unique structural, electronic, optoelectronic, semiconductor, mechanical, and chemical properties, making them a popular choice in wastewater treatment due to their high surface area and mesoporous nature (Biswas, 2020; Jin et al., 2018; Rashid et al., 2021; Wang, 2012). Modified CNTs, when functionalized with organic molecules covalently or non-covalently, exhibit even higher adsorption capacities (Biswas, 2020; Gupta et al., 2016). Researchers have studied the use of CNTs in removing various heavy metals such as Pb^{2+}, Cd^{2+}, and Cr^{2+} through adsorption using both modified and unmodified CNTs (Lu et al., 2017).

Graphene oxide (GO) is another advanced material studied extensively in recent years. Discovered in 2004, GO is a two-dimensional hexagonal lattice of carbon atoms and shares similar properties with CNTs (Biswas, 2020; Novoselov et al., 2012). GO has been used to adsorb Pb^{2+}, with a reported maximum adsorption capacity of 406.6 mg/g within 40 minutes (Deng et al., 2007). Another study showed that hydrogel lignosulphonate -modified GO could adsorb Pb^{2+} with a capacity of 1308 mg/g (Li et al., 2016). Additionally, the adsorption of other heavy metals, such as Cu^{2+}, Pb^{2+}, and Cd^{2+}, on magnetic GO has been studied, showing it to be an efficient adsorbent (Hur et al., 2015). The adsorption capacities of selected nano-adsorbents are summarized in Table 2.9.

TABLE 2.9

The adsorption capacity of selected nano-adsorbents for heavy metal ions and dye removal by adsorption

Adsorbent	Adsorbate	Adsorption capacity (mg/g)	References
Activated carbon (AC)	Crystal violet	84.11	Sarabadan et al. (2019)
Natural zeolite	Dye	177.75	Rashid et al. (2021)
MWCNTs	Reactive dyes	95–352.1	Machado et al. (2012)
Ni/PC-CNT	Congo red (CR)	818	Jin et al. (2018)
	Methylene blue	312	
	Malachite green	898	
JA-MIONs	Methylene blue	76.9	Ghosh et al. (2023)
	Antibiotics	200	
Magnetic bamboo-charcoal nanocomposite	Methylene blue	111.11	Sen (2023b)
Fe_3O_4 nanoparticles	Pd^{2+}	369.0–523.6	Xin et al. (2012)
	Cd^{2+}		
	Cu^{2+}		
TiO_2 nanotubes/CNT	Cu^{2+}	83–124	Sadegh et al. (2017)
	Pb^{2+}	192–588	
Multi-walled CNTs	Ni^{2+}	3.72	Yang et al. (2009)
CNTs	Cd^{2+}	144.48	Anitha et al. (2015)
CNTs-OH	Cd^{2+}	171.36	Anitha et al. (2015)
CNTS-$CONH_2$	Pb^{2+}		
PEI/GO	Pb^{2+}	64.94	Al-Yaari and Saleh (2023)

Various metal oxide nanomaterials, such as Fe_3O_4, CuO, ZnO, and their composites, have also been utilized in removing toxic metal ions and other pollutants from wastewater and raw water (Singh et al., 2018). Multiple nano-sized adsorbents have been synthesized and employed in water purification applications (Wang et al., 2015).

Bibliography

Abdolali, A., H. H. Ngo, W. Guo, S. Lu, S. S. Chen, N. C. Nguyen, X. Zhang, J. Wang, Y. Wu (2016). A breakthrough biosorbent in removing heavy metals: Equilibrium, kinetic, thermodynamic and mechanism analyses in a lab-scale study. *Science of the Total Environment*, 542, 603–611. https://doi.org/10.1016/j.scitotenv.2015.10.095

Abidi, N., E. Errais, J. Duplay, A. Berez, A. Jrad, G. Schäfer, M. Ghazi, K. Semhi, M. Trabelsi-Ayadi (2015). Treatment of dye-containing effluent by natural clay. *Journal of Cleaner Production*, 86, 432–440.

Afroza, S., T. K. Sen, M. Ang (2016a). Adsorption Performance of Continuous Fixed Bed Column for the removal of Methylene Blue (MB) Dye using Eucalyptus Sheathiana Bark biomass. *Research on Chemical Intermediates*, 42, 2343–2364. https://doi 10.1007/s11164-015-2153-8

Afroza, S., T. K. Sen, M. Ang (2016b). Adsorption removal of Zinc (II) from Aqueous Phase by Raw and Base Modified *Eucalyptus Sheathiana* Bark: kinetics, mechanism and equilibrium study. *Process Safety and Environmental Protection*, 102, 336–357. https://doi:10.1016/j.psep.2016.04.009

Afroze, S. (2016). Ph.D thesis entitled aqueous phase adsorption of organic/inorganic contaminants by eucalyptus bark (eucalyptus sheathiana) biomass. Curtin University, Perth, AUSTRALIA. URI: https://hdl.handle.net/20.500.11937/1561

Afroze, S., T. K. Sen (2018). Removal of aqueous phase heavy metal ions and dyes using agricultural solid wastes adsorbents: A review. *Water, Air & Soil Pollution*, 229(225), 1–50. https://doi.org/10.1007/s11270-018-3869-z

Ahmad, A. A., A. Idris, B. H. Hameed (2013). Organic dye adsorption on activated carbon derived from solid waste. *Desalinationn and Water Treatment*, 51, 2554–2563. https://doi.org/10.1080/19443994.2012.749019

Ahmad, A., Z. A. Ghazi, M. Saeed, M. Ilyas, R. Ahmad, A. Muqsit Khattak, A. Iqbal (2017). Comparative study of the removal of Cr(vi) from synthetic solution using natural biosorbents. *New Journal Chemistry*, 41, 10799–10807. https://doi.org/10.1039/C7NJ02026K

Ahmad, T., M. A. Danish (2022). Review of avocado waste-derived adsorbents: Characterizations, adsorption characteristics, and surface mechanism. *Chemosphere*, 296, 134036. https://doi.org/10.1016/j.chemosphere.2022.134036

Akram, M., H. N. Bhatti, M. Iqbal, S. Noreen, S. Sadaf (2017). Biocomposite efficiency *for Cr(VI)* adsorption: Kinetic, equilibrium and thermodynamics studies. *Jouranl of Environmental Chemical Engineering*, 5, 400–411. https://doi.org/10.1016/j.jece.2016.12.002

Aksu, Z. (2001). Equilibrium and kinetic modelling of cadmium(II) biosorption by C. vulgaris in a batch system: Effect of temperature. *Separation and Purification Technology*, 21, 285–294. https://doi.org/10.1016/S1383-5866(00)00212-4

AL- Yaari, M., T. A. Saleh (2023). Removal of Lead from Wastewater Using Synthesized Polyethyleneimine-Grafted Graphene Oxide. *Nanomaterials*, 13, 1078. https://doi.org/10.3390/nano13061078

Alarboday, R. S. A., F. Amran, M. A. A. Zaini (2024). Valorisation of cardboard waste adsorbents for efficient malachite green removal – Equilibrium, kinetics, and thermodynamics. *Energy Sources, Part A: Recovery, Utilization, and Environmental Effects*, 46(1), 5274–5289. https://doi.org/10.1080/15567036.2024.2336172

Ali, H. (2010). Biodegradation of synthetic dyes—A review. *Water, Air, & Soil Pollution*, 213(1), 251–273.

Anastopoulos, I., A. Bhatnagar, B. H. Hameed, Y. S. OK, M. Omirou (2017). A review on waste-derived adsorbents from sugar industry for pollutant removal in water and wastewater. *Journal Molecular Liquids*, 240, 179–188.

Angove, M. J., B. B. Johnson, J. D. Wells (1998). The influence of temperature on the adsorption of cadmium (II) and cobalt(II) on kaolinite. *Journal of Colloid Interface Science*, 204, 93–103.

Anitha, K., S. Namsani, J. K. Singh (2015). Removal of heavy metal ions using a functionalized single-walled carbon nanotube: A molecular dynamics study. *The Journal of Physical Chemistry A*, 119(30), 8349–8358.

Ansari, M., A. Aroujalian, A. Raisi, B. Dabir, M. Fathizadeh (2014). Preparation and characterization of nano-NaX Zeolite by microwave assisted Hydrothermal method. *Advanced Powder Technology*, 25(2), 722–727.

Ansari, S.A., F. Khan, A. Ahmad (2016). Cauliflower leave, an agricultural waste biomass adsorbent, and its application for the removal of MB dye from aqueous solution: Equilibrium, kinetics, and thermodynamic studies. *International Journal of Analytical Chemistry*, 2016, 1–10. https://doi.org/10.1155/2016/8252354

Apark, R., K. Guclu, M. H. Turgut (1998). Modeling of copper (II), cadmium (II), and lead (II) adsorption on red mud. *Journal of Colloid and Interface Science*, 203, 122–130.

Argun, M.E., D. Güclü, M. Karatas, M. (2014). Adsorption of Reactive Blue 114 dye by using a new adsorbent: Pomelo peel. *Journal of Industrial and Engineeing Chemistry*, 20, 1079–1084. https://doi.org/10.1016/j.jiec.2013.06.045

Arias, F., T. K. Sen (2009). Removal of zinc metal ion (Zn^{2+}) from its aqueous solution by kaolin clay minerals: a kinetic and equilibrium study. *Colloids and Surfaces A*, 348, 100–108.

Asadi, E., P. Haroonian, M. Ghaedi, A. Asfaram (2024). Removal of tartrazine and indigo carmine with layered double hydroxide-modified graphite nanocomposite: Isotherm, kinetics and response surface methodology (RSM) optimization. *Journal of Molecular Liquids*, 402, 124769. https://doi.org/10.1016/j.molliq.2024.124769

Asgher, M., H. N. Bhatti (2012). Evaluation of thermodynamics and effect of chemical treatments on sorption potential of (Citrus) waste biomass for removal of anionic dyes from aqueous solutions. *Ecological Engineering*, 38(1), 79–85.

Auta, M., B. Hameed (2012). Modified mesoporous clay adsorbent for adsorption isotherm and kinetics of methylene blue. *Chemical Engineering Journal*, 198–199, 219–227.

Ayob, S., N. Othman, W. Ali, H. Altowayti, F. S. Khalid, N. A. Bakar (2021). A Review on adsorption of heavy metals from wood-industrial wastewater by oil palm. *Journal of Ecological Engineering*, 22(3), 249–265. https://doi.org/10.12911/22998993/132854 ISSN 2299–8993, License CC-BY 4.0.

Banerjee, S., M. C. Chattopadhyaya (2017). Adsorption characteristics for the removal of a toxic dye, tartrazine from aqueous solutions by a low cost agricultural by-product. *Arabian Journal of Chemistry*, 10, S1629–S1638. https://doi.org/10.1016/j.arabjc.2013.06.005

Bansal, R. C., M. Goyal (2005). *Book on Activated Carbon Adsorption*. CRC Press, Boca Raton, USA, 1st ed., pages 520. Ebook ISBN: 9780429114182. https://doi.org/10.1201/9781420028812.

Barakat, M. A., E. Schmidt (2010). Polymer-enhanced ultrafiltration process for heavy metals removal from industrial wastewater. *Desalination*, 256, 90–93. https://doi.org/10.1016/j.desal.2010.02.008

Barka, N., M. Abdennouri, M. El Makhfouk, S. Qourzal (2013). Biosorption characteristics of cadmium and lead onto eco-friendly dried cactus (Opuntia ficus indica) cladodes. *Journal of Environmental Chemical Engineering*, 1, 144–149. https://doi.org/10.1016/j.jece.2013.04.0

Bayes, G. S., S. S. Raut, V. R. Patil, R. S. Lokhande (2012). Formation of diazepam-lanthanides(III) complexes in the 50-50 volume % ethanol-water solvent system and study of the effect of temperature on the complex formation constants. *Journal of Solution Chemistry*, 41, 241–248. https://doi.org/10.1007/s10953-012-9798-3

Bayuo, J., J. M. Rwiza, M. Sillanpa, K. M. Mte (2023). Removal of heavy metals from binary and multicomponent adsorption systems using various adsorbents – A systematic review. *RSC Advanced*, 13, 13052–13093. https://doi:10.1039/d3ra01660a

Bazan, A. P. Nowicki, R. Pietrzak (2016). Removal of NO2 by carbonaceous adsorbents obtained from residue after supercritical extraction of marigold. *Adsorption*, 22, 465–471. https://doi: 10.1007/s10450-015-9709-1

Ben-Ali, S., Jaouali, I., Souissi-Najar, S., Ouederni, A. (2017). Characterization and adsorption capacity of raw pomegranate peel biosorbent for copper removal. *Journal of Cleaner Production*, 142, 3809–3821. https://doi.org/10.1016/j.jclepro.2016.10.081

Benjamin, M. M., J. Leckie (1981). Multiplesite adsorption of Cd, Cu, Zn and Pb on amorphous iron oxyhydroxide. *Journal of Colloid and Interface Science*, 79, 209–219.

Bhatnagar A, M. Sillanpää (2010). Utilization of agro-industrial and municipal waste materials as potential adsorbents for water treatment— A review. *Chemical Engineering Journal*, 157, 277–296.

Bhatnagar, A., J. P. Vítor Vilar, M. S. Cidália Botelho, A. R. Rui Boaventura (2011). A review of the use of red mud as adsorbent for the removal of toxic pollutants from water and wastewater. *Environmental Technology*, 32, 231–249.

Bhatti, H. N., Q. Zaman, A. Kausar, S. Noreen, M. Iqbal (2016). Efficient remediation of Zr(IV) using citrus peel waste biomass: Kinetic, equilibrium and thermodynamic studies. *Ecological Engineering*, 95, 216–228. https://doi.org/10.1016 j.ecoleng.2016.06.08

Biswas, S. (2020). Ph.D thesis entitled Design and Development of a Three Phase Semifluidized Bed Reactor for Wastewater Treatment with Reference to Steel Plant Effluent. Curtin University, Perth, AUSTRALIA and IIT-Kharagpur, India under collaborative Ph.D. program. https://hdl.handle.net/20.500.11937/82127

Biswas, S., B. C. Meikap, T. K. Sen (2019a). Adsorptive removal of aqueous phase copper (Cu2+) and nickel (Ni2+) metal ions by synthesize Biochar-biopolymeric hybrid adsorbents and process optimization by response surface methodology (RSM), *Water, Air & Soil Pollution*, 230:197. https://doi.org/10.1007/s11270-019-4258-y

Biswas, S., H. Siddiqi, B. C. Meikap, T. K. Sen, M. Khiadani (2020a). Preparation and characterization of raw and inorganic acid-activated pine cone biochar and its application in the removal of aqueous-phase Pb^{2+} metal ions by adsorption. *Water, Air, and Soil Pollution*, 231(1), 3.

Biswas, S., Manisha Bal, S. K. Behera, T. K. Sen, B. C. Meikap (2019b). Process optimization study of Zn2+ adsorption on biochar-alginate composite adsorbent by Response Surface Methodology (RSM). *Water*, 11(2), 325. https://doi.org/10.3390/w11020325

Biswas, S., R. K. Diwakar, Ipshita D. Behera, B. C. Meikap, T. K. Sen, M. Khiadani (2020b). Aqueous phase phenol removal from synthetic and real steel plant effluents through a batch and Semifluidized bed clumn operation: Experimental and model analysis. *Journal Environmental Chemical Engineering*, 8, 104441. https://doi.org/10.1016/j.jece.2020.104441

Biswas, S., S. S. Mohapatra, U. Kumari, B. C. Meikap, T. K. Sen (2020c). Batch and continuous closed circuit semi-fluidized bed operation: Removal of MB dye using sugarcane bagasse biochar and alginate composite adsorbents. *Journal of Environmental Chemical Engineering*, 8(1), 103637. https://doi.org/10.1016/j.jece.2019.103637

Biswas, S., T. K. Sen, A. M. Yeneneh, B. C. Meikap (2019c). Synthesis and characterization of a novel Ca-alginate-biochar composite as efficient zinc (Zn^{2+}) adsorbent: Thermodynamics, process design, mass transfer and isotherm modelling. *Separation Science and Technology*, 54(7), 1106–1124.

Borah, L., M. Goswami, P. Phukan (2015). Adsorption of methylene blue and eosin yellow using porous carbon prepared from tea waste: Adsorption equilibrium, kinetics and thermodynamics study. *Journal of Environmental Chemical Engineering*, 3(2), 1018–1028.

Boruah, P., A. Sarma, K. G. Bhattacharyya (2015). Removal of Ni (II) ions from aqueous solution by using low cost biosorbent prepared from jackfruit (Artocarpus heterophyllus) leaf powder. *Indian Journal of Chemical Technology*, 22(6), 322–327.

Bozic, D., M. Gorgievski, V. Stankovic, N. Strbac, S. Serbula, N. Petrovic (2013). Adsorption of heavy metal ions by beech sawdust-kinetics, mechanism and equilibrium of the process. *Ecological Engineering*, 58, 502–206. https://doi.org/10.1016/j.ecoleng.2013.06.033

Brigatti, M.F., E. Galán, B.K.G. Theng (2013). Chapter 2 - Structure and Mineralogy of Clay Minerals". *Developments in Clay Science* Faïza, B. and Gerhard, L. (eds), Elsevier, pp. 21–81. Beijing. https://doi.org/10.1016/B978-0-08-098258-8.00002-X

Burakova, A. E., Evgeny V. G. I. V. Burakovaa, A. E. Kucherovaa, Shilpi Agarwalb, A. G. Tkacheva, V. K. Guptab (2018). Adsorption of heavy metals on conventional and nanostructured materials for wastewater treatment purposes: A review. *Ecotoxicology and Environmental Safety*, 148(2018), 702–707. https://doi.org/10.1016/j.ecoenv.2017.11.034

Chai, W., Y. Huang, G. Han, J. Liu, S. Yang, Y. Cao (2017). An enhanced study on adsorption of Al(iii) onto Bentonite and Kaolin: Kinetics, Isotherms, and Mechanisms. *Mineral Processing and Extractive Metallurgy Review*, 38(2), 106–115.

Chakraborty R, A. Asthana, A. K. Singh, B. Jain, A. B. H Susan (2020). Adsorption of heavy metal ions by various low-cost adsorbents: A review. *International Journal of Environmental Analytical Chemistry*, 102(2), 1–38. https://doi.org/10.1080/03067319.2020.1722811. Online version.

Chakrabortya, R., A. Asthanaa, A. K. Singha, B. Jaina, A. B. Susanb (2022). Adsorption of heavy metal ions by various low-cost adsorbents: A review. *International Journal of Environment Analytical Chemistry*, 102(2), 342–379. https://doi.org/10.1080/03067319.2020.1722811

Chen, H., J. Zhao, G. Dai, J. Wu, H. Yan (2010). Adsorption characteristics of Pb(II) from aqueous solution onto a natural biosorbent, fallen Cinnamomum camphora leaves. *Desalination*, 262(1–3), 174–182. https://doi.org/10.1016/j.desal.2010.06.006

Chen, L., Zhang, He-Xin (2022). Modified biochar adsorbents (MBAs) for heavy metal ions adsorption: A critical review. *Journal of Environmental Chemical Engineering*, 10(2), 107393.

Chen, Z., H. Deng, C. Chen, Y. Yang, H. Xu (2014). Biosorption of malachite green from aqueous solutions by Pleurotus ostreatus using Taguchi method. *Journal of Environmental Health Science and Engineeing*, 12, 63. https://doi.org/10.1186/2052-336X-12-63

Chieng, H. I., L. B. L. Lim, N. Priyantha (2015). Enhancing adsorption capacity of toxic malachite green dye through chemically modified breadnut peel: Equilibrium, thermodynamics, kinetics and regeneration studies. *Environmental Technology*, 36, 86–97. https://doi.org/10.1080/09593330.2014.938124

Choi, H.-J, S.-W. Yu, K. H. Kim (2016). Efficient Use of MgModified Zeolite in the Treatment of Aqueous Solution Contaminated with Heavy Metal Toxic Ions. *Journal of the Taiwan Institute of Chemical Engineers*, 63, 482–489.

Cornell, R. M., U. Schwertmann (1998). *The Iron Oxide Structure, properties, reactions, occurrences and uses.*, 1st edition. Wiley-VCH GmbH & Co. KGaA. ISBN 3-527-29669-7

Crini, G. (2006). Non-conventional low-cost adsorbents for dye removal: A review. *Bioresour ce Technology*, 97, 1061–1085. https://doi.org/10.1016/j.biortech.2005.05.001

Dağdelen, S., B. Acemioğlu, E. Baran, O. Koçer (2014). Removal of Remazol Brilliant Blue R From Aqueous Solution by Pirina Pretreated with Nitric Acid and Commercial Activated Carbon. *Water, Air, Soil Pollution*, 225, 1899. https://doi.org/10.1007/s11270-014-1899-8

Dahri, M.K., L.B. Lim, N. Priyantha, C. M. Chan (2016). Removal of Acid blue 25 using Cempedak Durian peel from aqueous medium: Isotherm, kinetics and thermodynamics studies. *International Food Research Journal*, 23(3), 1154–1163.

Das, B., S. Prakash, P. S. R. Reddy, V. N. Misra (2007). An overview of utilization of slag and sludge from steel industries, resources. *Conservation and Recycling*, 50, 40–57.

Davis, T. A., B. Volesky, A. Mucci (2003). A review of the biochemistry of heavy metal biosorption by brown algae. *Water Reserach*, 37, 4311–4330.

Dawood, S., T. K. Sen (2012). Removal of anionic dye Congo red from aqueous solution by raw pine and acid treated pinecone powder as adsorbent: Equilibrium, thermodynamic, kinetics, mechanism and process design. *Water Research*, 46, 1933–1946.

Dawood, S., T. K. Sen (2014). Review on dye removal from its aqueous solution into alternative cost-effective non-conventional adsorbents. *Journal and Chemical and Process Engineering*, 1, 1–7.

Dawood, S (2018). Synthesis and characterization of biomass and clay minerals based adsorbents for the removal of cationic dye and metal ion from wastewater by adsorption. *Ph.D. thesis*.2018, Curtin University Library, Bentley, 6845, AUSTRALIA. URI. https://hdl.handle.net/20.500.11937/69346

Dawood, S., T. Gupta, T. K. Sen (2017a). Adsorptive removal of methylene blue (MB) dye at kaolin clay-water interface: Kinetics, isotherm modelling and process design, Chapter-7 in Book entitled "*Clay Minerals: Properties, Occurrence and Uses*, edited by T. K. Sen, Nova Publishers Inc, New York. ISBN 978-1-53610-823-1, pp. 209–236.

Dawood, S., T. K. Sen, Chi Pan (2018). Performance and dynamic modelling of biochar and kaolin packed bed adsorption column for aqueous phase methylene blue (MB) dye removal. *Environmental Technology*, 40, 1–11. https://doi.org/10.1080/09593330.2018.1491065

Dawood, S., T. K. Sen, Chi Phan (2017b). Synthesis and Characterisation of Pine Cone bio-char in the removal of organic and inorganic pollutants from aqueous solution by adsorption: Kinetic, equilibrium, mechanism and thermodynamic. *Bioresource and Biotechnology*, 246, 76–78. https://doi.org/10.1016/j.biortech.2017.07.019

Dawood, Sara, T. K. Sen, Chi Phan (2014). Preparation and characterization of novel activated carbon from waste biomass pine cone and its application in Congo red dye adsorption from aqueous solution: Equilibrium, thermodynamic, kinetics, mechanism and process design. *Water, Air & Soil Pollution*, 225, 1818. https://doi.org/10.1007/s11270-013-1818-4

Deliyanni, E. A., G. Z. Kyzas, K. S. Triantafyllidis, K. A. Matis (2015). Activated carbons for the removal of heavy metal ions: A systematic review of recent literature focused on lead and arsenic ions. *Open Chemistry*, 13, 699–708. From the *journal Open Chemistry*. https://doi.org/10.1515/chem-2015-0087

Deng, L., Y. Su, H. Su, X. Wang, X. Zhu (2007). Sorption and Desorption of Lead (II) from Wastewater by Green Algae Cladophora Fascicularis. *Journal of Hazardous Materials*, 143(1–2), 220–225.

Deng, Z., K. Y. Fung, K. M. Ng, C. Wei (2016). Design of anaerobic fluidized bed bioreactor – Dyeing effluents. *Chemical Engineering Science*, 139(Supplement C), 273–284.

Deniz, F., S. Karaman (2011). Removal of Basic Red 46 dye from aqueous solution by pine tree leaves. *Chemical Engineering Journal*, 170(1), 67–74.

Dutta, B. K. (2007). *Principle of Mass Transfer and Separation Process*. PHI learning, India, 1st edition, ISBN-10: 8120329902, ISBN-13: 978–8120329904.

Dzombak, D. A., F. M. M. Morel (1990). *Surface Complexing Modelling. Hydrous Ferric Oxide*, John Wiley, 393–393.

Dzombak, D., F. M. M. Morel (1988). Sorption of cadmium on hydrous ferric oxide at high sorbate/sorbent ratios: Equilibrium, kinetics, and modelling. *Journal of Colloid and Interface Science*, 112, 589–597.

El-Baz, A. A., I. A. Hendyb, A. M. Dohdohc, M. I. Soru (2020). Adsorption technique for pollutants removal; current new trends and future challenges – A Review. *The Egyptian International Journal of Engineering Sciences and Technology*, 32, 1–24. https://eijest.journals.ekb.eg

El-Sayed, O. G. (2011). Removal of methylene blue and crystal violet from aqueous solutions by palm kernel fiber. *Desalination*, 272(1–3), 225–232.

Fu, F., Q. Wang (2011). Removal of heavy metal ions from wastewaters: A review. *Journal of Environmental Management*, 92(3), 407–418. https://doi.org/10.1016/j.jenvman.2010.11.011

Foo, K.Y., B. H. Hameed (2011). Microwave assisted preparation of activated carbon from pomelo skin for the removal of anionic and cationic dyes. *Chemical Engineering Journal*, 173(2), 385–390.

Gautam, R. K., A. Mudhoo, G. Lofrano, M. C. Chattopadhyaya (2014). Biomass-derived biosorbents for metal ions sequestration: Adsorbent modification and activation methods and adsorbent regeneration. *Journal of Environmental Chemical Engineering*, 2(1), 239–259.

Ge, J. C., S. K. Yoon, N. J. Choi (2018). Application of Fly Ash as an adsorbent for removal of air and water pollutants. *Applied Science*, 8, 1116, https://doi.org/10.3390/app8071116

Geankoplis, C. J. (2003). *Transport Processes and Separation Process Principles*, 4th edition. Prentice Hall. NJ, USA. ISBN 0-13-101367-X

Ghodbane, I., L. Nouri, O. Hamdaoui, M. Chiha (2008). Kinetic and equilibrium study for the sorption of cadmium(II) ions from aqueous phase by eucalyptus bark. *Journal of Hazardous Materials,* 152, 148–158. https://doi.org/10.1016/j.jhazmat.2007.06.079

Ghodbane, I., O. Hamdaoui (2008). Removal of mercury(II) from aqueous media using eucalyptus bark: Kinetic and equilibrium studies. *Journal of Hazardous. Materials,* 160, 301–309. https://doi.org/10.1016/j.jhazmat.2008.02.116

Ghosh, A., M. Ghosh Dastidar, T. Sreekrishnan (2015). Recent advances in bioremediation of heavy metals and metal complex dyes: Review. *Journal of Environmental Engineering,* 142(9), C4015003. https://doi.org/10.1061/(ASCE)EE.1943-7870.0000965

Ghosh, N., S. Sen · Goutam Biswas · Atul Saxena, P. K. Haldar (2023). Adsorption and desorption study of reusable magnetic iron oxide nanoparticles modifed with justicia adhatoda leaf extract for the removal of textile dye and antibiotic. *Water Air Soil Pollution,* 234, 202. https://doi.org/10.1007/s11270-023-06217-8

Gisi, S. D., G. Lofrano, M. Grassi, M. Notarnicola (2016). Characteristics and adsorption capacities of low-cost sorbents for wastewater treatment: A review. *Sustainable Materials and Technologies*, 9, 10–40.

Goswami, M., P. Phukan (2017). Enhanced adsorption of cationic dyes using sulfonic acid modified activated carbon. *Journal of Environmental Chemical Engineering,* 5(4), 3508–3551.

Guechi, E.-K., O. Hamdaoui (2016). Sorption of malachite green from aqueous solution by potato peel: Kinetics and equilibrium modeling using non-linear analysis method. *Arabian Journal of Chemistry*, 9, S416–S424. https://doi.org/10.1016/j.arabjc.2011.05.011

Gul, S., S. Gul, H. Gul, N. Raza, A. Azzouz, M. R. Elamin, L. Khezami (2024). Enhanced adsorptive removal of malachite green in environmental samples using Java plum leaves: From equilibrium to mechanism studies. *Biomass Conversion and Biorefnery*, 15, 4225–4241. https://doi.org/10.1007/s13399-024-05567-5

Gupta, V. (2009). Application of low-cost adsorbents for dye removal–A review. *Journal of Environmental Management, 90(8),* 2313–2342.

Gupta, V. K., A. Rastogi, A., Nayak (2010). Biosorption of nickel onto treated alga (Oedogonium hatei): Application of isotherm and kinetic models. *Journal of Colloid Interface Science*. 342, 533–539. https://doi.org/10.1016/j.jcis.2009.10.074

Gupta, V. K., O. Moradi, I. Tyagi, S. Agarwal, H. Sadegh, R. Shahryari-Ghoshekandi, A. S. H. Makhlouf, M. Goodarzi, A. Garshasbi.(2016). Study on the removal of heavy metal ions from industry waste by carbon nanotubes: Effect of the surface modification: A review. *Critical Reviews in Environmental Science and Technology,* 46(2), 93–118.

Gutha, Y., V. S. Munagapati, M. Naushad, K. Abburi (2015). Removal of Ni(II) from aqueous solution by Lycopersicum esculentum (Tomato) leaf powder as a low-cost biosorbent. *Desalination Water Treatment,* 54, 200–208. https://doi.org/10.1080/19443994.2014.880160

Guyo, U., M. Moyo (2017). Cowpea pod (Vigna unguiculata) biomass as a low-cost biosorbent for removal of Pb(II) ions from aqueous solution. *Environmental Monitoring and Assessment*, 189(2), 47.

Güzel, F., H. Sayğılı, G. Akkaya Sayğılı, F. Koyuncu, C. Yılmaz (2017). Optimal oxidation with nitric acid of biochar derived from pyrolysis of weeds and its application in removal of hazardous dye methylene blue from aqueous solution. *Journal of Cleaner Production*, 144(Supplement C), 260–265.
Hamdaoui, O. (2017). Adsorption of Cu(II) from aqueous phase by Cedar bark. *Journal of Dispersion Science* and *Technology*, 38, 1087–1091. https://doi.org/10.1080/01932691.2016.1225261
Han, X., J. Yuan, X. Ma (2014). Adsorption of malachite green from aqueous solutions onto lotus leaf: Equilibrium, kinetic, and thermodynamic studies. *Desalination and Water Treatment*, 52, 5563–5574. https://doi.org/10.1080/19443994.2013.813102
Harris, D. C. (2010). Summary for Policymakers, in: *Intergovernmental Panel on Climate Change (Ed.), Climate Change 2013- The Physical Science Basis.* Cambridge University Press, Cambridge, pp. 1–30. https://doi.org/10.1017/CBO9781107415324.004
Ho, Y. S., C. T. Huang, H. W. Huang (2002). Equilibrium sorption isotherm for metal ions on tree fern. *Process Biochemistry*, 37(12), 1421–1430.
Hor, K. Y., J. M. C. Chee, M. N. Chong, B. Jin, C. Saint, P. E. Poh, R. Aryal (2016). Evaluation of physicochemical methods in enhancing the adsorption performance of natural zeolite as low-cost adsorbent of methylene blue dye from wastewater. *Journal of Cleaner Production*, 118(Supplement C), 197–209.
Hur, J., J. Shin, J. Yoo, Y.-S. Seo (2015). Competitive adsorption of metals onto magnetic graphene oxide: Comparison with Other carbonaceous adsorbents. *The Scientific World Journal*, 2015, 1–11.
Hussain, S. A., S. Demirci, G. Ozbayoglu (1996). Zeta potential measurements on three clays from turkey and effects of clays on coal flotation. *Journal of Colloid and Interface Science*, 184, 535–541.
Hussin, Z. M., N. Talib, N. M. Hussin, M. A. Hanafiah, W. K. Khalir (2015). Methylene blue adsorption onto NaOH modified durian leaf powder: Isotherm and kinetic studies. *American Journal of Environmental Engineering*, 5(3A), 38–43.
Ibrahim, R. I. (2020). Optimization process for removing of copper ions from groundwater of Iraq. Using watermelon shell as natural adsorbent. IOP Conference series: *Materials Sci Engineering*, 737, 012195. https://doi.org/10.1088/1757-899X/737/1/012195
Iijima, S. (1991). Helical Microtubules of Graphitic Carbon. *Nature*, 354(6348), 56–58.
Isahak, W. N. R. W., Hisham, M. W. M., Yarmo, M. A. (2012). Highly porous carbon materials from biomass by chemical and carbonization method: A comparison study. *Journal of Chemistry*, 2013, 1–6. https://doi.org/10.1155/2013/620346
Jain, A. K., V. K. Gupta, A. Bhatnagar, Suhas (2003). Utilization of industrial waste products as adsorbents for the removal of dyes. *Journal of Hazardous Materials B*, 101, 31–42.
Jin, L., X. Zhao, X. Qian, M. Dong (2018). Nickel nanoparticles encapsulated in porous carbon and carbon nanotube hybrids from bimetallic metal-organic-frameworks for highly efficient adsorption of dyes. *Jornal of Colloid and Interface Science*, 509, 245–253. https://doi.org/10.1016/j.jcis.2017.09.002
Joh, D., J. K. Mbadcam (2015). Adsorption of Zinc (II) ions from aqueous solution onto kaolinite and metakaolinite. *Der Pharm Chemica*, 7, 51–58.

Kaithwasa, A., M. Prasadb, A. Kulshreshthab, S. Vermaa (2012). Industrial wastes derived solid adsorbents for CO2 capture: A mini review. *Chemical Engineering Research and Design,* 90, 1632–1641. https://doi.org/10.1016/j.cherd.2012.02.011

Kamari, S., A. Shahbazi (2020). Biocompatible Fe3O4@SiO2-NH2 nanocomposite as a green nanofiller embedded in PES–nanofiltration membrane matrix for salts, heavy metal ion and dye removal: Long–term operation and reusability tests. *Chemosphere,* 243, 125282. https://doi.org/10.1016/j.chemosphere.2019.125282

Kansal, S. K., A. Kumari (2014). Potential of M. Oleifera for the Treatment of Water and Wastewater. *Chemical Reviews,* 114(9), 4993–5010.

Kaplan, D. (2013). Absorption and Adsorption of Heavy Metals by Microalgae, in: *Handbook of Microalgal Culture.* Oxford, John Wiley & Sons, Ltd, pp. 602–611.

Kara, İ., D. Yilmazer, S. T. Akar (2017). Metakaolin based geopolymer as an effective adsorbent for adsorption of zinc(II) and nickel(II) ions from aqueous solutions. *Applied Clay Science,* 139, 54–63.

Karamveer, S. S., D. Kapoor, N. Singh, A. K. Saini, W. F. Alsanie, V. K. Thakur (2022). Air pollutants removal using biofiltration technique: A challenge at the frontiers of sustainable environment. *ACS Engineering Au,* 2022, 2(5), 378–396. https://doi.org/10.1021/acsengineeringau.2c00020

Khalid, K., W. S. Ngah, M. A. Hanafiah, N. S. Malek, N.S., S. N. Khazaai (2015). Acid Blue 25 adsorption onto phosphoric acid treated rubber leaf powder. *American Journal of Environmental Engineering,* 5(3A), 19–25.

Khan, T. A., E. A. Khan, Shahjahan (2015). Removal of basic dyes from aqueous solution by adsorption onto binary iron-manganese oxide coated kaolinite: Non-linear isotherm and kinetics modeling. *Applied Clay Science,* 107(Supplement C), 70–77.

Khilar, K. C., H. S. Fogler (1998). *Migration of Fines in Porous Media. Chapter 1 and Chapter 4,* 1st edition, Kluwer Academic Publishers.

Kılıç, M., C. Kırbıyık, O. Çepelioğullar, A. E. Pütün (2013). Adsorption of heavy metal ions from aqueous solutions by biochar, a by-product of pyrolysis. *Applied Surface Science,* 283, 856–862.

Kołodyńska, D., J. Krukowska, P. Thomas (2017a). Comparison of sorption and desorption studies of heavy metal ions from biochar and commercial active carbon. *Chemical Engineering Journal,* 307(Supplement C), 353–363.

Kołodyńska, D., J. Krukowska-Bąk, J. Kazmierczak-Razna, R. Pietrzak (2017b). Uptake of heavy metal ions from aqueous solutions by sorbents obtained from the spent ion exchange resins. *Microporous and Mesoporous Materials,* 244(Supplement C), 127–136.

Krishna, L. S., A. Yuzir, G., Yuvaraja, V. Ashokkumar (2017). Removal of Acid Blue25 from aqueous solutions using Bengal gram fruit shell (BGFS) biomass. *International Journal of Phytoremediation,* 19(5), 431–438.

Krishnan, K. A., K. Sreejalekshmi, V. Vimexen, V. V. Dev (2016). Evaluation of adsorption properties of sulphurised activated carbon for the effective and economically viable removal of Zn (II) from aqueous solutions. *Ecotoxicology and Eenvironmental Ssafety,* 124, 418–425.

Kuila T, S. Bose, A. K. Mishra, P. Khanra, N. H. Kim, J. H. Lee (2012). Chemical functionalization of graphene and its applications. *Progress in Materials Science,* 57, 1061–1105.

Kumar, P. S., S. Ramalingam, S. D. Kirupha, A. Murugesan, T. Vidhyadevi, S. Sivanesan (2011). Adsorption behavior of nickel(II) onto cashew nut shell: Equilibrium, thermodynamics, kinetics, mechanism and process design. *Chemical Engineering Journal*, 167, 122–131. https://doi.org/10.1016/j.cej.2010.12.010

Lafi, R., A. Fradj, A. Hafiane, B. H. Hameed (2014). Coffee waste as potential adsorbent for the removal of basic dyes from aqueous solution. *Korean Journal of Chemical Engineering*, 31(12), 1–9. https://doi.org/10.1007/s11814-014-0171-7

Lamai, C., M. Kruatrachue, P. Pokethitiyook, E. S. Upatham, V. Soonthornsarathool (2005). Toxicity and accumulation of lead and cadmium in the filamentous green alga cladophora fracta (O.F. Muller Ex Vahl) Kutzing : A laboratory study. *Science Asia*, 31, 121–127.

Lee, K. M., C. W. Lai, K. S. Ngai, J. C. Juan (2016a). Recent developments of zinc oxide based photocatalyst in water treatment technology: A review. *Water Research*, 88, 428–448.

Lee, K.M., Y. M. Jo (2010). Synthesis of zeolite from waste fly ash for adsorption of CO2. *Journal of Material Cycles and Waste Management*, 12, 212–219.

Lee, L. Y., S. Gan, M. S. Yin Tan, S. S. Lim, X. J. Lee, Y. F. Lam (2016b). Effective removal of Acid Blue 113 dye using overripe Cucumis sativus peel as an eco-friendly biosorbent from agricultural residue. *Journal of Cleaner Production*, 113(Supplement C), 194–203.

Li, Y., A. Meas, S. Shan, R. Yang, X. Gai (2016). Production and optimization of bamboo hydrochars for adsorption of Congo red and 2-naphthol. *Bioresource Technology*, 207(Supplement C), 379–386.

Li, Fengfeng, X. Wang, T. Yuan, R. Sun (2016). A lignosulfonateModified graphene hydrogel with ultrahigh adsorption capacity for Pb(ii) removal. *Journal of Materials Chemistry A*, 4(30), 11888–11896.

Liang, F. B., Y. L. Song, C. P. Huang, J. Zhang, B. H. Chen (2013). Adsorption of hexavalent chromium on a lignin-based resin: Equilibrium, thermodynamics, and kinetics. *Journal of Environmental Chemical Engineering*, 1(4), 1301–1308. https://doi.org/10.1016/j.jece.2013.09.025

Liu, Q., Y. Zhou J. Lu Zhou (2020). Novel cyclodextrin-based adsorbents for removing pollutants from wastewater: A critical review. *Chemosphere*, 241, 125043.

Liu, Y., L. Liu, Y. Wang (2021). A critical review on removal of gaseous pollutants using sulfate radical-based advanced oxidation techniques. *Environmental Science and Technology*, 55(14), 9691–9710. https://doi.org/10.1021/acs.est.1c0153

Lothenbach, B., G. Furrer, R. Schulin (1997). Immobilization of heavy metals by polynuclear aluminum and montmorillonite compounds. *Environmental Science and Technology*, 31, 1452–1462.

Lu, W., J. Li, Y. Sheng, X. Zhang, J. You, J. Chen (2017). One-pot synthesis of magnetic iron oxide nanoparticle-multiwalled carbon nanotube composites for enhanced removal of Cr(VI) from aqueous solution. *Journal of Colloid and Interface Science*, 505, 1134–1146. https://doi.org/10.1016/j.jcis.2017.07.013

Machado F. M, C. P. Bergmann, T. H. Fernandes, E. C. Lima, B. Royer, T. Calvete, S. B. Fagan (2011). Adsorption of Reactive Red M-2BE dye from water solutions by multi-walled carbon nanotubes and activated carbon. *Journal of Hazardous Materials*, 192, 1122–1131.

Magdy, A., Y. O. Fouad, M. H. Abdel-Aziz, A. H. Konsowa (2017). Synthesis and characterization of Fe3O4/kaolin magnetic nanocomposite and its application in wastewater treatment. *Journal of Industrial and Engineering Chemistry*, 56(Supplement C), 299–311.

Mahmoodi, N. M., B. Hayati, M. Arami, C. Lan (2011). Adsorption of textile dyes on Pine Cone from colored wastewater: Kinetic, equilibrium and thermodynamic studies. *Desalination*, 268(1–3), 117–125.

Malik, R., S. Dahiya (2017). An experimental and quantum chemical study of removal of utmostly quantified heavy metals in wastewater using coconut husk: A novel approach to mechanism. *International Journal of Biological Macromolecules*, 98, 139–149. https://doi.org/10.1016/j.ijbiomac.2017.01.100

Mall, I., V. Srivastava, G. Kumar, I. Mishra (2006). Characterization and utilization of mesoporous fertilizer plant waste carbon for adsorptive removal of dyes from aqueous solution. *Colloids and Surfaces A: Physicochemical and Engineering Aspects*, 278(1), 175–187.

Mallampati, R., L. Xuanjun, A. Adin, S. Valiyaveettil (2015). Fruit peels as efficient renewable adsorbents for removal of dissolved heavy metals and dyes from water. *ACS Sustainable Chemistry and Engineerng*, 3, 1117–1124.

Marrakchi, F., M. J. Ahmed, W. A. Khanday, M. Asif, B. H. Hameed (2017). Mesoporous-activated carbon prepared from chitosan flakes via single-step sodium hydroxide activation for the adsorption of methylene blue. *International Journal of Biological Macromolecules*, 98, 233–239.

Mckenzie, R. M. (1980). The adsorption of lead and other heavy metals on oxides of manganise and iron. *Australian Journal of Soil Research*, 18, 61–73.

Miraboutalebi, S. M., S. K. Nikouzad, M. Peydayesh, N. Allahgholi, L. Vafajoo, G. McKay.(2017). Methylene blue adsorption via maize silk powder: Kinetic, equilibrium, thermodynamic studies and residual error analysis. *Process Safety and Environmental Protection*, 106, 191–202. https://doi.org/10.1016/j.psep.2017.01.010

Mohamed, E. F., A. El-Mekawy, S. A. Sayed Ahmed, N. A. Fathy (2024). High Adsorption Capacity of Ammonia Gas Pollutant Using Adsorbents of Carbon Composites. *Arabian Journal for Science and Engineering*, 49, 261–271.

Mohameda, T., L. Nadjiab, D. Abdelkadera, E. Abdelkader, B. abderrahmanea, Z. Billal (2024). Characterization and applicability of the Naima–Tiaret–Algeria illitemontmorillonite clay for methylene blue dye removal: Adsorption kinetic, isotherm and thermodynamic studies. *Water Practice & Technology*, 19(5), 1726–1752. https://doi.org/10.2166/wpt.2024.077

Mohammed, M., T. K. Sen, S. Maitra, B. K. Dutta (2011). Removal of Zn^{2+} from aqueous solution using Castor seed hull. *Water Air and Soil Pollution*, 215 (1–4), 609–620. https://doi.org/10.1007/s11270-010-0503-0

Monteiro, C. M., P. M. L. Castro, F. X. Malcata (2010). Cadmium removal by two strains of desmodesmus pleiomorphus cells. *Water Air, & Soil Pollution*, 208, 17–27. https://doi.org/10.1007/s11270-009-0146-1

Mukherjee, K., A. Kedia, K. Jagajjanani Rao, S. Dhir, S. Paria (2015). Adsorption enhancement of methylene blue dye at kaolinite clay-water interface influenced by electrolyte solutions. *RSC Advances*, 5(39), 30654–30659.

Nawaz, S., H. N. Bhatti, T. H. Bokhari, S. Sadaf (2014). Removal of Novacron Golden Yellow dye from aqueous solutions by low-cost agricultural waste: Batch and fixed bed study. *Chemistry and Ecology*, 30, 52–65. https://doi.org/10.1080/02757540.2013.841898

Neubaker, U., B. Nowalk, G. Furrer, R. Schulin (2000). Heavy mwtal sorption on clay minerals affected by the siderophore desferrixamine B. *Environmental Science and Technology*, 34, 2749–2755.

Nishioka, H., T. K. Sen (2019). Solvothermal synthesis and characterization of magnetic bamboo charcoal (BC) nanocomposites. *Journal of Institution of Engineers (India) Series E*, 100, 155–165.

Nomura, S., H. Nishioka, T. K. Sen (2017). Synthesis and arsenic adsorptive characteristics of a novel magnetic adsorbent. *Journal of Environmental Conservation Engineering*, 46(3), 35–42.

Novoselov, K. S., V. I. Fal'ko, L. Colombo, P. R. Gellert, M. G. Schwab, K. Kim (2012). A Roadmap for graphene. *Nature*, 490(7419), 192–200.

Nowicki1, P. J., P. Kazmierczak-Raznal, 1. Skibiszewska, M. Wis´niewska, A. Nosal-Wiercin´ska, R. Pietrzak (2016). Production of activated carbons from biodegradable waste materials as an alternative way of their utilisation. *Adsorption*, 22, 489–502. https://doi.org/ 10.1007/s10450-015-9719-z

Nowicki, P., H. Wachowska, R. Pietrzak (2010). Active carbons prepared by chemical activation of plum stones and their application in removal of NO_2. *Journal of Hazardous Materis*. 181, 1088–1094.

Ogunlalu, O., I. P. Oyekunle, O. Kingsley, B. C. lwuozor, A. D. Aderibigbe, E. C. Emenike (2021). Trends in the mitigation of heavy metal ions from aqueous solutions using unmodified and chemically modified agricultural waste adsorbents. *Current Research in Green and Sustainable Chemistry*, 4, 100188.

Önal, E., N. Özbay, A. Ş. Yargıç, R. Z. Y. Şahin, Ö. Gök (2014). Performance Evaluation of the Bio-Char Heavy Metal Removal Produced from Tomato Factory Waste, in: *Progress in Exergy, Energy, and the Environment*. Springer International Publishing, Cham, pp. 733–740. ISBN: 978-3-319-04680 https://doi.org/10.1007/978-3-319-04681-5_70

Öztürk, A., E. Malkoc (2015). Cationic Basic Yellow 2 (BY2) adsorption onto manure ash: surface properties and adsorption mechanism. Desalination *Water Treatment*, 54, 209–226. https://doi.org/10.1080/19443994.2013.878671

Paranavithana, G. N., K. Kawamoto, Y. Inoue, T. Saito, M. Vithanage, C. S. Kalpage, G. B. B. Herath (2016). Adsorption of Cd2+ and Pb2+ onto coconut shell biochar and biochar-mixed soil. *Environmental Earth Sciences*, 75, 484. https://doi.org/10.1007/s12665-015-5167-z

Park, G. A (1967). *Aqueous Surface Chemistry of Oxides and Complex Oxide minerals Isoelectric Point and Zero Point of Charge in Equilibrium Concepts in Natural Water Systems*, (W.stumm edi), Advances in Chemistry, 67, American chemical society, Washington DC. ISBN: 9780841200685, eISBN: 9780841222472

Pavan, F. A., E. S. Camacho, E. C. Lima, G. L. Dotto, V. T. Branco, S. L. Dias (2014). Formosa papaya seed powder (FPSP): Preparation, characterization and application as an alternative adsorbent for the removal of crystal violet from aqueous phase. *Journal of Environmental Chemical Engineering*, 2(1), 230–238.

Peng, J., Q. Xiao, Z. Wang, F. Zhou, J. Yu, R. Chi, C. Xiao (2024). Mechanistic investigation of Pb2+ adsorption on biochar modifed with sodium alginate composite zeolitic imidazolate framework 8. *Environmental Science and Pollution Research*, 31, 31605–31618. https://doi.org/10.1007/s11356-024-33320-y

Pesol, M. F. M., N. Amri, F. Senusi, S. F. A. Halim, M. S. M. Sidik (2024). Modification of the coffee industry sludge-based adsorbent using acid treatment for the removal of tartrazine dye from aqueous solution. *Esteem Academic Journal*, 20, 82–97. https://doi.org/10.24191/esteem.v20iMarch. 580.g5 36

Pietrzak, R. (2010). Sawdust pellets from coniferous species as adsorbents for NO_2 removal. *Bioresource Technology*, 101, 907–913.
Purkait, M. K., A. Maiti, S. DasGupta, S. De (2007). Removal of congo red using activated carbon and its regeneration. *Journal of Hazardous Materials*, 145(1–2), 287–295.
Putra, W. P., A. Kamari, S. N. M. Yusoff, C. F. Ishak, A. Mohamed, N. Hashim, N. I. M. Isa (2014). Biosorption of Cu (II), Pb (II) and Zn (II) ions from aqueous solutions using selected waste materials. *Journal of Encapsulation and Adsorption Sciences*, 4(1), 741. https://doi.org/10.4236/jeas.2014.41004
Qing Du, Z. Sun, W. Forslins, H. Tang (1997). Adsorption of copper at aqueous illite surfaces. *Journal of Colloid and Interface Science*, 187, 232–242.
Qiu, G., Q. Xie, H. Liu, T. Chen, J. Xie, H. Li (2015). Removal of Cu (II) from aqueous solutions using dolomite–Palygorskite clay: Performance and mechanisms. *Applied Clay Science*, 118, 107–115.
Rafatullah, M., O. Sulaiman, R. Hashim, A. Ahmad (2010). "Adsorption of methylene blue on low-cost adsorbents: A review. *Journal of Hazardous Materials*, 177(1–3), 70–80.
Rambabu, K., F. Banat, Q. M. Pham, S. H. Ho, N. Q. Ren, P. L. Show (2020). Biological remediation of acid mine drainage: Review of past trends and current outlook. *Environmental Science Ecotechnology*, 2, Article 100024.
Randhawa, N. S., N. N. Das, R. K. Jana (2014). Adsorptive remediation of Cu(II) and Cd(II) contaminated water using manganese nodule leaching residue. *Desalin Water Treatment* 52, 4197–4211. https://doi.org/10.1080/19443994.2013.801324
Rangsayatorn, N., E. S. Upatham, M. Kruatrachue, P. Pokethitiyook, G. R. Lanza (2002). Phytoremediation potential of Spirulina (Arthrospira) platensis: Biosorption and toxicity studies of cadmium. *Environmental Pollution*, 119, 45–53. https://doi.org/10.1016/S0269-7491(01)00324-4
Rashid, R., I. Shafiq, P. Akhter, M. J. Iqbal, M. Hussain (2021). A state-of-the-art review on wastewater treatment techniques: The effectiveness of adsorption method. *Environmental Science and Pollution Research*, 28, 9050–9066. https://doi.org/10.1007/s11356-021-12395-x/
Ren, X., W. Xiao, R. Zhang, Y. Shang, R. Han (2013). Adsorption of crystal violet from aqueous solution by chemically modified phoenix tree leaves in batch mode. *Desalin Water Treatment*, 53(5), 1–11. https://doi.org/10.1080/19443994.2013.859105
Rida, K., S. Bouraoui, S. Hadnine (2013a). Adsorption of methylene blue from aqueous solution by kaolin and zeolite. *Applied Clay Science*, 83–84(Supplement C), 99–105.
Rida, K., S. Bouraoui, S. Hadnine (2013b). Adsorption of methylene blue from aqueous solution by kaolin and zeolite. *Applied Clay Science*, 83–84, 99–105.
Sadaf, S., H. N. Bhatti, S. Ali, K.-U. Rehman (2014). Removal of Indosol Turquoise FBL dye from aqueous solution by bagasse, a low cost agricultural waste: Batch and column study. *Desalination and Water Treatment*, 52(1–3), 184–198.
Sadegh H, G. A. Ali, V. K. Gupta, A. S. H. Makhlouf, R. Shahryari-ghoshekandi, M. N. Nadagouda, M. Sillanpää, E. Megiel (2017). The role of nanomaterials as effective adsorbents and their applications in wastewater treatment. *Journal of Nanostructure in Chemistry*, 7, 1–14.
Saha, R., K. Mukherjee, I. Saha, A. Ghosh, S. K. Ghosh, B. Saha (2013). Removal of hexavalent chromium from water by adsorption on mosambi (Citrus limetta) peel. *Research on Chemical Intermediates*, 39, 2245–2257. https://doi.org/10.1007/s11164-012-0754-z

Said, A. E.-A. A., A. A. M. Aly, M. N. Goda, M. Abd El-Aal, M. Abdelazim (2017). Modified Sugarcane Bagasse with Tartaric Acid for Removal of Diazonium Blue from Aqueous Solutions. *Journal of Polymers and the Environment*, 26, 2424–2433. https://doi.org/10.1007/s10924-017-1136-9

Salleh, M. A. M., D. K. Mahmoud, W. A. Karim, A. Idris (2011). Cationic and anionic dye adsorption by agricultural solid wastes: A comprehensive review. *Desalination*, 280(1–3), 1–13.

Samonin, V. V., V. Yu. Nikonova, M. L. Podvyaznikov (2014). Carbon adsorbents on the basis of the hydrolytic lignin modified with fullerenes in producing. *Russian Journal of Applied Chemistry*, 87(2), 190–193.

Sangi, M. R., A. Shahmoradi, J. Zolgharnein, G. H. Azimi, M. Ghorbandoost (2008). Removal and recovery of heavy metals from aqueous solution using Ulmus carpinifolia and Fraxinus excelsior tree leaves. *Journal Hazardous Materials*, 155, 513–522. https://doi.org/10.1016/j.jhazmat.2007.11.110

Sarabadan, M., H. Bashiri, S. M. Mousavi (2019). Removal of crystal violet dye by an efficient and low cost adsorbent: Modeling, kinetic, equilibrium and thermodynamic studies. *Korean Journal of Chemical Engineering*, 36, 1575–1586.

Saravanan, R., A. K. Priya, P. Senthil Kumar, K. Tuan, A. Hoang, K. Sekar (2022). A critical and recent developments on adsorption technique for removal of heavy metals from wastewater-A review *Chemosphere*, 303(2), 135146.

Sarma, G. K., S. Sen Gupta, K. G. Bhattacharyya (2016). Adsorption of Crystal violet on raw and acid-treated montmorillonite, K10, in aqueous suspension. *Journal of Environmental Management*, 171, 1–10.

Shaukat, M.I., R. Wahi, Z. Ngaini (2022). The application of agricultural wastes for heavy metals adsorption: A meta-analysis of recent studies. *Bioresource Technology Reports*, 100902. https://doi.org/10.1016/j.biteb.2021.100902

Sayğılı, G. A., F. Güzel (2017). Chemical modification of a cellulose-based material to improve its adsorption capacity for anionic dyes. *Journal of Dispersion Science and Technology*, 38(3), 381–392.

Schroth, B. K., G. Sposito (1998). Effect of landfill leachate organic acids on trace metal adsorption by kaolinite. *Environmental Science and Technology*, 32, 1404–1408.

Scida, K., P. W. Stege, G. Haby, G. A. Messina, C. D. García (2011). Recent applications of carbon-based nanomaterials in analytical chemistry: Critical review. *Analytica Chimica Acta*, 691(1–2), 6–17.

Seader, J. D., E. J. Henley (1998). *Separation Process Principles*. Wiley.

Sen, T. K. (2001). Ph. D. thesis entitled Studies on Colloidal fines-associated contaminant transport in saturated porous media at Indian Institute of Technology, Bombay, India. Call no. 043:541.182:628.54Sen, Barcode 194744. https://opac.library.iitb.ac.in/cgi-bin/koha/opac-detail.pl?biblionumber=209599

Sen, T. K. (2015). *Physical, Chemical and Biological Treatment Processes for Water and Wastewater*, Nova Publishers Inc., 1st edition, 2015, ISBN: 978-1-63483-396-7. (https://www.novapublishers.com/catalog/product_info.php?products_id=55622).

Sen, T. K. (2017). *Air, Gas and Water Pollution Control Using Industrial and Agricultural Solid Wastes Adsorbents*. CRC-Press, Taylor & Francis, 1st edition, ISBN: 13:978[[sbn]]1[[sbn]]138[[sbn]]19673[[sbn]]5 (hhttps://www.crcpress.com/Air-Gas-and-Water-Pollution-Control-Using-Industrial-and Agricultural/Sen/p/book/9781138196735).

Sen, T. K. (2023a). Agricultural solid wastes based adsorbent materials in the remediation of heavy metal ions from water and wastewater by adsorption: A review. *Molecules*, 28(14), 555. https://doi.org/10.3390/molecules28145575

Sen, T. K. (2023b). Application of synthesized biomass bamboo charcoal–iron oxide "BC/Fe" nanocomposite adsorbents in the removal of cationic methylene blue dye contaminants from wastewater by adsorption. *Sustainability*, 15, 8841. https://doi.org/10.3390/su15118841

Sen, T. K., S. Afroze, H. M. Ang (2011). Equilibrium, Kinetics and Mechanism of Removal of Methylene Blue from Aqueous Solution by Adsorption onto Pine Cone Biomass of Pinus radiata. *Water, Air, Soil Pollution*, 218, 499–515. https://doi.org/10.1007/s11270-010-0663-y

Sen, T. K., Amir Farid Bin Azman, S. Maitra, B. K. Dutta (2011). Removal of mercury (II) by Leaves of Rambai Tree. *Water Environment Research*, 83(9), 834–842.

Sen, T. K., M. Verlyny Sarzali (2008). *Removal of cadmium metal ion (Cd2+) from its aqueous solution by aluminium oxide (Al2 O3),* A kinetic and equilibrium study. *Chemical Engineering Journal*, 142, 256–262.

Sen, T., K. Dustin Gomez (2011). Adsorption of zinc from aqueous solution on natural bentonite. *Desalination*, 267(2–3), 286–294. https://doi.org/10.1016/j.desal.2010.09.041

Senthil Kumar, P., M. Palaniyappan, M. Priyadharshini, A. Vignesh, A. Thanjiappan, P. Sebastina Anne Fernando, R. Tanvir Ahmed, R. Srinath (2014). Adsorption of basic dye onto raw and surface-modified agricultural waste. *Environmental Progress & Sustainable Energy*, 33(1), 87–98.

SenthilKumar, P., S. Ramalingam, V. Sathyaselvabala, S. D. Kirupha, S. Sivanesan (2011). Removal of copper(II) ions from aqueous solution by adsorption using cashew nut shell. *Desalination*, 266, 63–71. https://doi.org/10.1016/j.desal.2010.08.003

Sewu, D. D., P. Boakye, H. Jung, H. S. H. Woo (2017a). Synergistic dye adsorption by biochar from co-pyrolysis of spent mushroom substrate and Saccharina japonica. *Bioresource Technology*, 244(Part 1), 1142–1149.

Sewu, D. D., P. Boakye, P. S. H. Woo (2017b). Highly efficient adsorption of cationic dye by biochar produced with Korean cabbage waste. *Bioresource Technology*, 224, 206–213.

Shaban, M., M. R. Abukhadra, M. Shahien, S. S. Ibrahim, S.S. (2017). Novel bentonite/zeoliteNaP composite efficiently removes methylene blue and Congo red dyes. *Environmental Chemistry Letters*, 1–6. https://doi.org/10.1007/s10311-017-0658-7

Shaheen, S. M., A. S. Derbalah, F. S. Moghanm (2012). Removal of heavy metals from aqueous solution by zeolite in competitive sorption system. *International Journal of Environmental Science and Development*, 3, 362–367. https://doi.org/10.7763/IJESD.2012.V3.248

Shaikh, R. B., B. Saifullah, F. Rehman (2018). Greener method for the removal of toxic metal ions from the wastewater by application of agricultural waste as an adsorbent. *Water*, 10(10), 1316. https://doi.org/10.3390/w10101316

Shan, T. C., M. A. Matar, E. A. Makky, E. N. Ali (2017). The use of Moringa oleifera seed as a natural coagulant for wastewater treatment and heavy metals removal. *Applied Water Science*, 7(3), 1369–1376.

Sharma, Y. C., V. Srivastava, V. Singh, S. Kaul, C. Weng (2009). Nanoadsorbents for the removal of metallic pollutants from water and wastewater. *Environmental Technology*, 30, 583–609.

Sharma, N., D. Tiwari, S. Singh (2014). The efficiency appraisal for removal of malachite green by potato peel and neem bark: Isotherm and kinetic studies. *International Journal of Chemical & Environmental Engineering*, 5(2), 84–88.

Shaukat, M. I., R. Wahi, Z. Ngaini (2022). The application of agricultural wastes for heavy metals adsorption: A meta-analysis of recent studies. *Bioresource Technology Reports*, 17, 100902. https://doi.org/10.1016/j.biteb.2021.100902

Shukla, S. R., R. S. Pai (2005). Adsorption of Cu(II), Ni(II) and Zn(II) on dye loaded groundnut shells and sawdust. *Separation Purification Technology*, 43, 1–8. https://doi.org/10.1016/j.seppur.2004.09.003

Singh, N., G. Nagpal, S. Agrawal (2018).Water purification by using adsorbents: A review. *Environment Technology and Innovation*, 11, 187–240.

Sivalingam, S., A. Sowmiya (2024). Industrial biomass waste as an economical, potential adsorbent for removing the Bismarck Brown R dye and zinc metal ions from effluents. *Environmental Science Advances*.3, 732–750. https://doi.org/10.1039/d3va00314k

Sivalingam, S., V. Gopal (2024). Low-cost adsorbent from biomass for removal of Fe(II) and Mn(II) for water treatment: Batch and column adsorption study. *Chemical Papers*, https://doi.org/10.1007/s11696-024-03438-x

Smith, S. C., D. F Rodrigues (2015). Carbon-based nanomaterials for removal of chemical and biological contaminants from water: A review of mechanisms and applications. *Carbon*, 91, 122–143.

Sočo, E., J. Kalembkiewicz (2013). Adsorption of nickel(II) and copper(II) ions from aqueous solution by coal fly ash. *Journal of Environmental Chemical Engineering*, 1(3), 581–588.

Soliman, N. K., A. F. Moustafa (2020). Industrial solid waste for heavy metals adsorption features and challenges: a review. *Journal of Materials Research and Technology*, 9(5), 10235–10253. https://doi.org/10.1016/j.jmrt.2020.07.045

Song, S.-T., N. Saman, K. Johari, H. Mat (2014). Surface chemistry modifications of rice husk toward enhancement of Hg(II) adsorption from aqueous solution. *Clean Technologies and Environmental Policy*, 16(8), 1747–1755. https://doi.org/10.1007/s10098-014-0803-y

Srivastava, S., S. B. Agrawal, M. K. Mondal (2015). Biosorption isotherms and kinetics on removal of Cr(VI) ssing native and chemically Modified Lagerstroemia Speciosa Bark. *Ecological Engineering*, 85, 56–66.

Stavropoulos, G., G. Skodras, K. Papadimitriou (2015). Effect of solution chemistry on cyanide adsorption in activated carbon. *Applied Thermal Engineering*, 74, 182–185.

Sulyman, M., J. Namiesnik, A. Gierak (2017). A. Low-cost Adsorbents Derived from Agricultural By-products/Wastes for Enhancing Contaminant Uptakes from Wastewater: A Review. *Polish Journal of Environmental Studies*, 26(2), 479–510. Review. https://doi.org/10.15244/pjoes/66769

Tamura, H., R. Furuichi (1997). Adsorption affinity of divalent heavy metal ions for metal oxides evaluated by modelling with the Frumkin isotherm. *Journal of Colloid and Interface Science*, 195, 241–249.

Tanweer, A., M. Danish (2022). A review of avocado waste-derived adsorbents: Characterizations, adsorption characteristics, and surface mechanism. *Chemosphere*, 296, 134036. https://doi.org/10.1016/j.chemosphere.2022.134036. 2020.

Tee, Guat T, W. F. Gok, Xie Yuen, W. F, Yong (2022). Review article Adsorption of pollutants inwastewater via biosorbents, nanoparticles and magnetic biosorbents: A review. *Environmental Research*, 212 (B), 113248. https://doi.org/10.1016/j.envres.2022.113248

Thirumavalavan, M., Y.-L. Lai, L.-C. Lin, J.-F. Lee (2010). Cellulose-based native and surface modified fruit peels for the adsorption of heavy metal ions from aqueous solution: Langmuir adsorption isotherms. *Journal of Chemical and Engineering Data,* 55, 1186–1192. https://doi.org/10.1021/je900585t

Tiadi N, R. Dash, C. Mohanty, A. Patel (2020). Comparative studies of adsorption of chromium (VI) ions onto different industrial wastes. *Journal of Hazardous Toxic and Radioactive Waste,* 24, 04020021.

Tien, C. (1994). *Adsorption Calculation and Modelling,* 1st edition. Butterworth-Heinemann, ISBN 0-7506-9121-2.

Tina Co Nnie Tan, T. K. Sen (2019). Thermodynamics, equilibrium and kinetic studies of aqueous phase methylene blue (MB) dye adsorption onto mixture of eucalyptus bark biomass and kaolin clay minerals. *Separation Science and Technology,* 1–15. https://doi.org/10.1080/01496395.2019.1580734

Trivedi, P., L. Axe (2000). Modelling Cd and Zn sorption to hydrous metal oxides. *Environmental Science Technology*, 34, 2215–2223.

Tuzen, M., A. Sari, D. Mendil, O. D. Uluozlu, M. Soylak, M. Dogan (2009). Characterization of biosorption process of As(III) on green algae Ulothrix cylindricum. *Journal of Hazardous Materials*, 165, 566–572. https://doi.org/10.1016/j.jhazmat.2008.10.020

Uçar, S., M. Erdem, T. Tay, S. Karagöz (2015). Removal of lead (II) and nickel (II) ions from aqueous solution using activated carbon prepared from rapeseed oil cake by Na2CO3 activation. *Clean Technologies and Environmental Policy,* 17(3), 747–756.

Uddin, M. T., M. A. Rahman, M. Rukanuzzaman, M. A. Islam (2017). A potential low cost adsorbent for the removal of cationic dyes from aqueous solutions. *Applied Water Science*, 7(6), 2831–2842.

Vaġzoğullar, A. Ġ., M. Uğurlu, G. Kula (2015). Comparing adsorption activity of raw Sepiolite and CTAB modified Sepiolite: Kinetic and adsorption study for removal of Hg^{2+}. *International Journal of Environment,* 4(4), 19–31.

Valcárcel, M., S. Cárdenas, B. M. Simonet, Y. Moliner-Martínez, R. Lucena (2008). Carbon Nanostructures as Sorbent Materials in Analytical Processes. *TrAC Trends in Analytical Chemistry,* 27(1), 34–43.

Volesky, B. (2007). Biosorption and me. *Water Research,* 41(18), 4017–4029.

Vulava, V. M., R. Kretzschmar, U. Rusch, D. Grolimund, J. C. Westall, M. Borkovec (2000). Cation competition in a natural subsurface material: Modeling of sorption equilibria. *Environmental Science Technology,* 34, 2149–2155.

Wang, H., X. Yuan, Z. Wu, L. Wang, X. Peng, L. Leng, G. Zeng (2014). Removal of basic dye from aqueous solution using cinnamomum camphora sawdust: Kinetics, isotherms, Thermodynamics, and mass-transfer processes. *Separation Science and Technology*, 49, 2689–2699. https://doi.org/10.1080/01496395.2014.940590

Wang, W, T. Jiao, Q. Zhang, X. Luo, J. Hu, Y. Chen, Q. Peng, X. Yan, B. Li (2015). Hydrothermal synthesis of hierarchical core–shell manganese oxide nanocomposites as efficient dye adsorbents for wastewater treatment. *RSC Advances,* 5, 56279–56285.

Wang, X. (2012). Nanomaterials as sorbents to remove heavy metal ions in wastewater treatment. *Journal of Environmental & Analytical Toxicology,* 02(07), 2–7. ISSN: 2161-0525. https://doi.org/10.4172/2161-0525.1000154

Wei, M., Q. Yu, T. Mu, L. Hou, Z. Zuo, Jiayan Peng (2016). Preparation and characterization of waste ion-exchange resinbased activated carbon for CO2 capture. *Adsorption*, 22, 385–396. https://doi.org/10.1007/s10450-016-9787-8

Wu, Y., Y. Fan, M. Zhang, Z. Ming, S. Yang, A. Arkin, P. Fang (2016). Functionalized agricultural biomass as a low-cost adsorbent: Utilization of rice straw incorporated with amine groups for the adsorption of Cr(VI) and Ni(II) from single and binary systems. *Biochemical Engineering Journal*, 105(Part A), 27–35.

Xin X, Q. Wei, J. Yang, L. Yan, R. Feng, G. Chen, B. Du, H. Li (2012). Highly efficient removal of heavy metal ions by aminefunctionalized mesoporous Fe3O4 nanoparticles. *Chemical Engineering Journal*, 184, 132–140.

Yadav, V. S., M. Prasad, J. Khan, S. S. Amritphale, M. Singh, C. B. Raju (2010). Sequestration of carbon dioxide (CO2) using red mud. *Journal of Hazardous Materials*, 176, 1044–1050.

Yagub, Mustafa, T. K. Sen, S. Afroza, H. A. Ming (2014). Dye and its removal from aqueous solution by Adsorption: A review. *Advances in Colloids and Interface Science*, 209, 172–184.

Yang, S., J. Li, D. Shao, J. Hu, X. Wang (2009). Adsorption of Ni (II) on oxidized multi-walled carbon nanotubes: Effect of contact time, pH, foreign ions and PAA. *Journal of Hazardous Materials*, 166, 109–116.

Yao, Y., H. Zuo, Y. Liu, S. Pang, L. Lan, F. Yao, Y. Wuab, Z. Li (2024). Efficient dye adsorption of mesoporous activated carbon from bamboo parenchyma cells by phosphoric acid activation. *RSC Advances*, 14, 1287.

Yargic, A. S., R. Z. Yarbay, N. Sahin, O. E. Onal (2015). Assessment of toxic copper(II) biosorption from aqueous solution by chemically-treated tomato waste. *Journal of Cleaner Production*, 88, 152–159. https://doi.org/10.1016/j.jclepro.2014.05.087

Yildiz, H. (2024). The production of a novel adsorbent from forest waste (Platanus orientalis L.) for dye adsorption: Adsorption process optimization and experimental design. *Materials Science & Engineering B*, 304, 117366. https://doi.org/10.1016/j.mseb.2024.117366

Yin, C. Y., M. K. Aroua, W. M. A. W. Daud (2007). Impregnation of palm shell activated carbon with polyethyleneimine and its effects on Cd2+ adsorption. *Colloids and Surfaces A: Physicochemical and Engineering Aspects*, 307(1–3), 128–136.

Zazycki, M. A., M. Godinho, D. Perondi, E, L. Foletto, G. C. Collazzo, G. L. Dotto (2018). New biochar from pecan nutshells as an alternative adsorbent for removing reactive red 141 from aqueous solutions. *Journal of Cleaner Production*, 171(Supplement C), 57–65.

Zhang, Z., I. M. O'Hara, G. A. Kent, W. O. S. Doherty (2013). Comparative study on adsorption of two cationic dyes by milled sugarcane bagasse. *Indstrial Crops and Products*, 42, 41–49. https://doi.org/10.1016/j.indcrop.2012.05.008

Zhang, Y., H. Ou, H. Liu, Y. Ke, W. Zhang, G. Liao, D. Wang (2018). Polyimide based carbon nanofibers: A versatile adsorbent for highly efficient removals of chlorophenols, dyes and antibiotics. *Colloids and Surfaces A*, 537, 92–101.

Zhao, B., Y. Shang, W. Xiao, C. Dou, R. Han (2014). Adsorption of Congo red from solution using cationic surfactant modified wheat straw in column model. *Jouranl of Environmental Chemical Engineering*, 2(1), 40–45. https://doi.org/10.1016/j.jece.2013.11.025

Zhao, Z., B. Wang, B. K. G. Theng, X. Lee, X. Zhang, M. Chen, P. Xu (2022). Removal performance, mechanisms and influencing factors of biochar for air pollutants: a critical review. *Biochar*, 4, 30. https://doi.org/10.1007/s42773-022-00156-z

Zhou, N., H. Chen, J. Xi, D. Yao, Z. Zhou, Y. Tian, X. Lu (2017). Biochars with excellent Pb(II) adsorption property produced from fresh and dehydrated banana peels via hydrothermal carbonization. *Bioresource Technology*, 232, 204–210.

Zhu, L., Y. Wang, T. He, L. You, X. Shen, X. (2016). Assessment of potential capability of water bamboo leaves on the adsorption removal efficiency of cationic dye from aqueous solutions. *Journal of Polymers and the Environment*, 24(2), 148–158.

Zwain, H. M., M. Vakili, I. Dahlan (2014). Waste material adsorbents for zinc removal from wastewater: A comprehensive review. *International Journal of Chemical Engineering*, 2014, 1–13. https://doi.org/10.1155/2014/347912

3

Adsorption Theory, Kinetics and Mechanistic Models, Mode of Operation, Design Calculations and Application

3.1 Introduction

The adsorption process involves a solid phase (adsorbent) and a bulk liquid or gaseous phase containing the dissolved species to be adsorbed (adsorbate), which accumulates on the solid adsorbent surface. This process continues until adsorption equilibrium is established between the amount of solid-bound sorbate species, q_e (mg/g), and its residual concentration remaining in the solution, C_e (mg/L) (Kratochvil & Volesky, 1998). Equilibrium occurs when the concentration of the sorbate/solute in the bulk solution is in dynamic balance with that at the interface, meaning the rate of adsorption equals the rate of desorption (Afroze & Sen, 2018; Oladoja et al., 2008).

There are two modes of operation in the adsorption process: batch and continuous. Batch adsorption experiments provide critical information about the effectiveness of the adsorbent and help determine various system parameters and their optimal values. However, the data obtained from batch studies are not directly applicable to continuous systems, where contact times are typically shorter than equilibrium times. Consequently, continuous adsorption systems are more accurate in determining the maximum adsorption capacity and adsorbent utilization. Such systems are crucial for evaluating the feasibility of adsorbents on an industrial scale (Afroze & Sen, 2018; Dawood, 2018).

To design processes, control adsorption phenomena, and identify adsorption mechanisms under varying operating parameters, quantifying the adsorption capacities of different selective adsorbents for removing water and gaseous pollutants is essential. This can be achieved through batch equilibrium and kinetic experimental studies.

DOI: 10.1201/9781003373223-3

3.2 Adsorption Theory, Models and Calculations, and applications

3.2.1 Adsorption Equilibria and Calculation of Adsorbed Amount

To determine the amount of adsorption at equilibrium, batch adsorption experiments are conducted in a stirred tank reactor, as shown in Figure 3.1. The procedure is as follows:

1. Fill a reaction vessel with a fixed volume, V (L), of a solution with an initial adsorbate concentration, C_0 (mg/L). Add m(g) of the adsorbent in pellet or powder form, ensuring the mixture is well-stirred (Figure 3.1). For gaseous-phase adsorbates, the concentration, C, is replaced by partial pressure, p.
2. Maintain specific physicochemical conditions (e.g., temperature, solution pH, adsorbent dose, and stirrer speed) and allow sufficient time for equilibrium to be reached, where the rate of adsorption equals the rate of desorption.
3. After equilibrium is achieved, sample and filter the slurry solution, then analyse the concentration of the adsorbate adsorbed.

Laboratory batch adsorption equilibrium experiments are conducted using a known amount of adsorbent with a fixed volume of aqueous solutions containing varying initial adsorbate concentrations. These experiments are

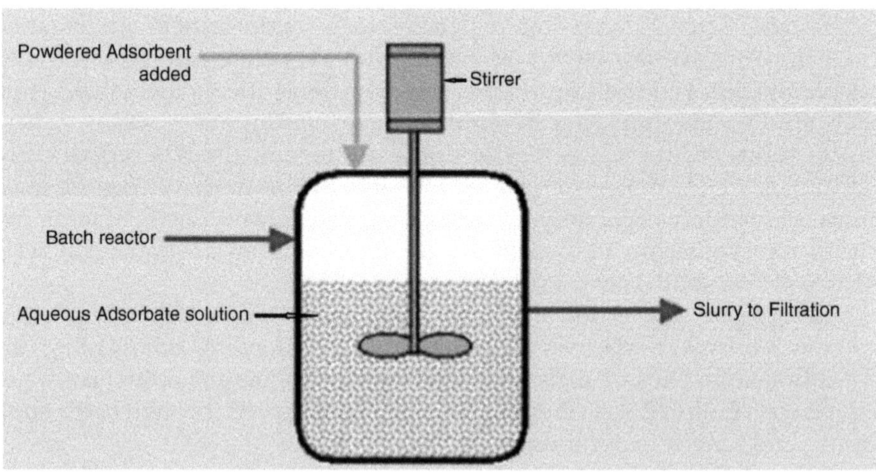

FIGURE 3.1
A stirred tank reactor for adsorption studies

FIGURE 3.2
Picture of Thermoline Scientific orbital shaker incubator

Source: Afroze, 2016a.

performed in a series of reaction vessels (e.g., plastic bottles) placed on a constant-temperature shaker, as shown in Figure 3.2. The reaction time should exceed the equilibrium time, which is determined through batch adsorption kinetic experiments. After the reaction, the samples are separated, filtered, and analysed to determine the remaining adsorbate concentration in the solution at equilibrium (C_e, mg/L).

The mass balance for adsorbate in the reaction bottle can be expressed as:

$$V(C_0 - C_e) = m(q_e - q_0) \tag{3.1}$$

where

- C_0 = initial adsorbate concentration (mg/L)
- C_e = equilibrium adsorbate concentration (mg/L)
- q_e = solid-phase adsorbate concentration at equilibrium (mg/g)
- q_0 = initial solid-phase adsorbate concentration (mg/g)
- V = volume of reaction mixture (L)
- m = amount of adsorbent added into the system (g).

If $q_0 = 0$, Eq. 3.1 simplifies to:

$$q = \frac{(C_0 - C_e)V}{m} \tag{3.2}$$

The percentage removal of adsorption is calculated as:

$$\%\ \text{removal of adsorption} = \frac{C_0 - C_e}{C_0} \times 100 \tag{3.3}$$

The equilibrium relationship between the amount of adsorption q (mass of solute adsorbed per unit mass of adsorbent, mg/g) and the equilibrium solute concentration C_e (or partial pressure p, for gaseous-phase solutes), at a constant temperature T, is known as the adsorption isotherm (Dutta, 2007):

$$q = f(Ce) \tag{3.4}$$

This functional relationship depends on the temperature. A plot of the equilibrium adsorbent loading q_e vs. temperature T at constant pressure is referred to as an isobar (Dutta, 2007; Kovach, 1997).

Various types of adsorption isotherms exist for different adsorbent-—adsorbate systems (Dutta, 2007). These isotherms illustrate the relationship between the amount of adsorbate adsorbed on the surface of an adsorbent and the pressure at a constant temperature (Figure 3.3).

Braunauer et al. (1940) identified five basic types of adsorption isotherms (Figure 3.4):

Liu et al. (2024) reviewed various two-parameter and three-parameter kinetic and isotherm model equations for biochar (BCs) and modified biochar (MBCs) in the sorption of toxic metal ions (TMIs) from solutions

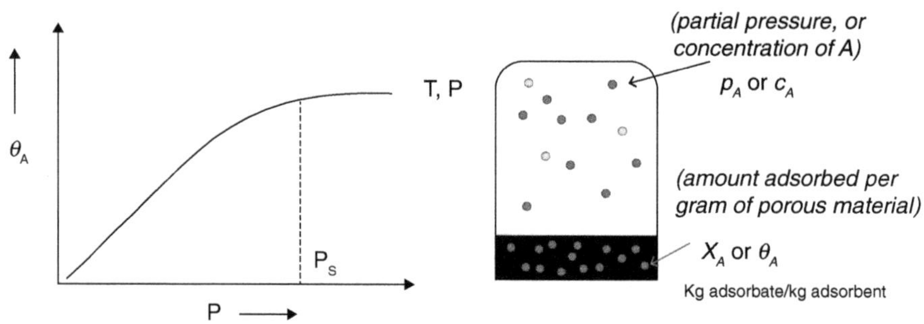

FIGURE 3.3
A graph obtained from experimental data for a specific adsorbent/adsorbate pair under —specified conditions (Courtesy: Seader and Henley, 1998)

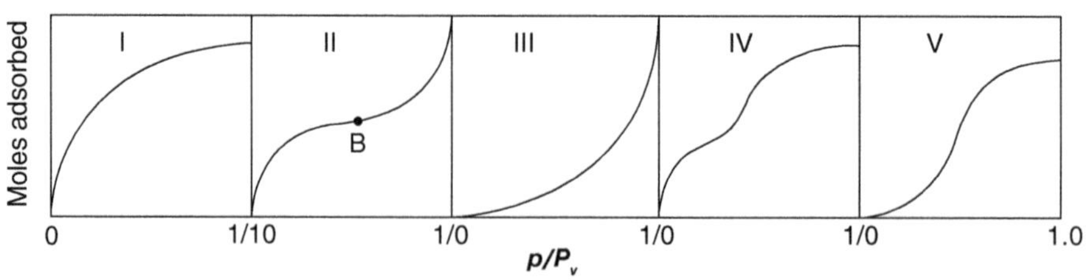

FIGURE 3.4
Five basic types of adsorption isotherms identified by Braunauer et al. (1940). Here, p = —adsorbate pressure and p_v = saturation vapour pressure of the adsorbate

containing single metal ions, mixed metal ions, and mixed metal ions/organics systems. The two-parameter isotherm models include Langmuir, Freundlich, BET, Dubinin–Radushkevich (D–R), Temkin, Harkins–Jura (H–J), and Halsey isotherm equations, while the three-parameter adsorption models are Redlich–Peterson, Sips, and Toth models (Liu et al., 2024). The two-parameter isotherm model equations are generally obtained via the linear fitting of experimental data using the linear form of these isotherm equations, whereas the three-parameter models are derived through the non-linear fitting of data. A few of the two-parameter and three-parameter isotherm model equations, along with their applications in solid/liquid and gas/solid adsorption systems, are presented below.

3.2.1.1 Type I Langmuir Adsorption Isotherm

This type of adsorption isotherm is concave downwards (Figure 3.4) and can be easily explained by Langmuir-type adsorption. Monolayer adsorption of adsorbate takes place on the adsorbent surface, as depicted in Figure 3.5. This adsorption technique is typically applied in the removal of solutes present in low concentrations from water or gaseous phases. For example, the adsorption of nitrogen (N_2) or hydrogen (H_2) on charcoal at –180°C follows this type of adsorption isotherm.

3.2.1.2 Langmuir Adsorption Isotherm

The Langmuir isotherm (1918) is based on the simplest model of the physical dynamic adsorption process, which assumes that:

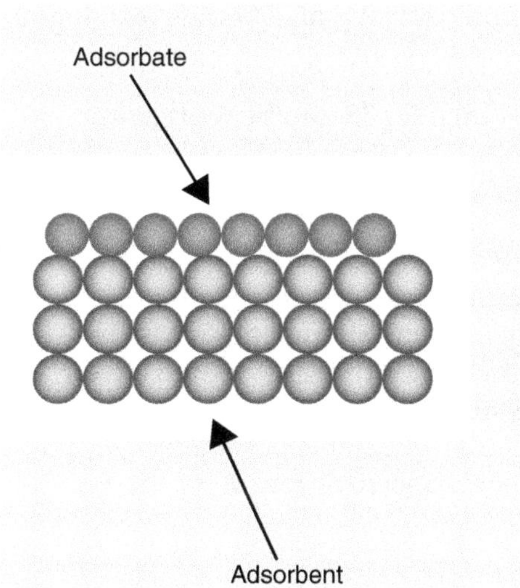

FIGURE 3.5
Monolayer adsorption

(i) The surface of the adsorbent is in contact with a solution containing an adsorbate that is strongly attracted to the surface.
(ii) The molecules are adsorbed at discrete active sites on the surface.
(iii) The adsorption involves the attachment of only one layer of molecules to the surface, i.e., monolayer adsorption (Febrianto, Kosasih et al., 2009; Ghaedi, Najibi et al., 2012).
(iv) The adsorbing surface is energetically uniform (Afroze & Sen, 2018).

For single-solute adsorption at the solid/liquid interface, the Langmuir isotherm indicates how the solute adsorbate molecules distribute between the liquid and solid phases at equilibrium during the adsorption process. At equilibrium, the number of molecules being adsorbed will be equal to the number of molecules leaving the adsorbed state through desorption. Let θ be the fraction of the adsorbent surface sites covered by solute adsorbate molecules.

At dynamic equilibrium:
Rate of adsorption = Rate of desorption.

$$\text{i.e.}\quad k_1 C_e(1-\theta) = k_2\theta \tag{3.5}$$

$$\text{or}\quad \theta = \frac{k_1 C_e}{k_2 + k_1 C_e} = \frac{KC_e}{1+KC_e} \tag{3.6}$$

where C_e is the concentration of adsorbate solute at equilibrium (mg/L), and

$K = \frac{k_1}{k_2}$ is equilibrium constant, also known as Langmuir constant.

If we write, $\theta = \frac{q_e}{q_m}$, then Eq 3.6 can be written as:

$$q_e = \frac{q_m K C_e}{1+K C_e} \tag{3.7}$$

where q_e is the amount of adsorbate adsorbed at equilibrium time (mg/g), C_e is the equilibrium concentration of adsorbate in solution (mg/L), q_m is the maximum adsorption capacity (mg/g) at high adsorbate concentrations, and K is the Langmuir isotherm constant (L/mg). Eq. 3.7 represents the Langmuir isotherm, which has two adjustable parameters: q_m and K.

At a low adsorbate concentration, $K\ C_e$ in Eq. 3.7 << 1, then Eq. 3.7 simplifies to:

$$q_e = K C_e\, q_m = K_H\, C_e \tag{3.8}$$

Eq. 3.8 is in the form of *Henry's law* and K_H is called Henry's law constant for adsorption (Dutta, 2007).

At high adsorbate concentration ($K\,C_e >> 1$), $q = q_m$ and fractional coverage $\theta = 1$, then the two different linearized forms of Langmuir isotherm equation I and Langmuir isotherm equation (II) can be written as (Afroze & Sen, 2018):

$$\frac{c_e}{q_e} = \frac{c_e}{q_m} + \frac{1}{Kq_m} \tag{3.9}$$

If this Langmuir I model is applicable to a particular adsorption system, a plot of the batch isotherm experimental data in the form of C_e/q_e vs. C_e will produce a straight line with the slope of $1/q_m$ and an intercept of $1/K\,q_m$. Therefore, the two adjustable isotherm parameters, q_m and K, can be found from the slope and intercept of the fitted straight line with the experimental data.

Similarly, the linearized Langmuir II isotherm model equation can be written as:

$$\frac{1}{q_e} = \frac{1}{Kq_m}\frac{1}{C_e} + \frac{1}{q_m} \tag{3.10}$$

If this Langmuir II model is applicable to a particular adsorption system, a plot of the batch isotherm experimental data in the form of $\frac{1}{q_e}$ vs. $\frac{1}{C_e}$ will produce a straight line with the slope of $\frac{1}{Kq_m}$ and an intercept of $\frac{1}{q_m}$ and with high linear regression coefficient (R^2) value. Therefore, the two adjustable isotherm parameters, q_m and K, can be found from the slope and intercept of the fitted straight line with the experimental data.

The separation factor (R_L) from Langmuir model can be calculated as per Eq. 3.11:

$$R_L = \frac{1}{1 + (K\,C_0)} \tag{3.11}$$

The separation factor (R_L) is a dimensionless unit used to determine the feasibility of an adsorption system (Kaur et al., 2013), as presented in Table 3.1 for various initial adsorbate concentrations C_0 (mg/L). Generally, the applicability of the isotherm model is chosen based on the higher linear regression coefficient (R^2) value during the fitting of the linearized model equation to the batch adsorption isotherm experimental data set. Table 3.2 presents the literature information on the applicability of the Langmuir isotherm model

TABLE 3.1

Feasibility of adsorption process at various (R_L) values Kaur et al. (2013)

Separation equilibrium factor (R_L)	Feasibility of adsorption process
$R_L > 1$	Unfavourable
$R_L = 1$	Linear
$0 < R_L < 1$	Favourable
$R_L = 1$	Irreversible

equation for different adsorbate-adsorbent systems, including both single and mixed competitive adsorption systems.

To determine the various model parameters for isotherm studies, both the linear models (Eq. 3.9 or Eq. 3.10) and the non-linear models (Eq. 3.7) must be fitted with the experimental batch adsorption isotherm data set. Many researchers have found that the model parameter values are sometimes more accurate when using non-linear model fitting with minimal error functions compared to linear fitting with the experimental data (Afroze, 2016a; Biswas, 2020d; Dawood, 2018).

Most studies in the adsorption field focus on the sequestration of heavy metals from single solute systems, but real wastewater often contains multiple components. Therefore, multicomponent adsorption must be studied to understand the competitive effects of the various inorganic and organic pollutants as well as the interactions between different pollutants (Bayuo et al., 2023). Multicomponent adsorption isotherm models are essential for establishing relationships between the uptake capacity of one pollutant by the adsorbent and the equilibrium concentration of other coexisting pollutants in real wastewater (Bayuo et al., 2023).

The simplest form of the unmodified Langmuir isotherm model, given by Eq. 3.7, can be adapted for multicomponent adsorption, as presented in Eq. 3.12 (Bayuo et al., 2023). Sometimes, the extended modified Langmuir isotherm model, given by Eq. 3.13, is applicable to multicomponent competitive adsorption systems (Das et al., 2022; Dutta, 2007). For example, in mixed metal ion sorption within multicomponent solution systems, interactions among different metal ions must be considered. Therefore, the amount of each component adsorbed at equilibrium in a multicomponent competitive adsorption system will be lower and affected by the presence of other competitive adsorbate metal ions. Consequently, the modified Langmuir isotherm model from Eq. 3.7 can be rewritten as Eq. 3.13 for multicomponent adsorption systems (Das, 2022; Dutta, 2007).

$$q_{e,i} = \frac{q_{\max,i} K_{L,i} C_{e,i}}{1 + \sum_{j=1}^{N} \left(K_{L,j} C_{e,j} \right)} \tag{3.12}$$

$$q_{e,i} = \frac{q_{m,i} K_{L,i} \left(\frac{c_{e,i}}{\eta_{L,i}} \right)}{1 + \sum_{N}^{j=1} K_{L,j} \left(\frac{c_{e,j}}{\eta_{L,j}} \right)} \tag{3.13}$$

where K_L is the modified Langmuir parameter. $\eta_{L,j}$ is the interactive parameter of the '*i*' th component in competitive adsorption system and '*N*' is the quantity of the tested data points for multicomponent solution. $C_{e,j}$ and $q_{e,i}$ are the equilibrium concentration and the adsorbed amount of the '*i*' component per unit gram of adsorbent, respectively. $C_{e,j}$ is the equilibrium concentration of jth component in mixed solution. $q_{\max,i}$ is monolayer adsorption capacity for component *i* (mg g–1) and $K_{L,i}$ and $K_{L,j}$ are the Langmuir constants obtained from the single component Langmuir isotherm model.

Table 3.2 presents very few recent studies reporting information on the applicability of Langmuir isotherm models for various single and multiple water and gaseous pollutant removal by various adsorbents.

TABLE 3.2

Recent studies on the applicability of Langmuir isotherm models for removal of water and gaseous pollutants by various adsorbents

Adsorbent	Water or gaseous pollutant adsorbates	Applicability of Langmuir isotherm	References
Acid-modified aloevera plant	Mn^{2+}, Fe^{3+}	Applicable	Sivalingam and Gopal (2024)
Date seed-based activated carbon	Zn^{2+}	Applicable	Sivalingam and Sowmiya (2024)
Modified orange peels	Cu^{2+}	Applicable	Chakraborty et al. (2022)
Biochar-alginate hybrid	Zn^{2+}	Applicable	Biswas et al. (2019a)
Ca-alginate biochar	Zn^{2+}	Applicable	Biswas et al. (2019b)
Biochar-biopolymeric hybrid	Cu^{2+}, Ni^{2+}	Applicable	Biswas et al. (2019c)
Acid-activated pinecone biochar	Pb^{2+}	Applicable	Biswas et al. (2020a)
Sugarcane bagasse	Hg^{2+}	Applicable	Chakraborty et al. (2020)
Rice straw-derived biochar	Binary Pb^{2+} & Zn^{2+}	Applicable	Sakhiya et al. (2022)
Passion peels	Binary Pb^{2+}, Cr^{3+}	Applicable	Castaneda-Fig et al. (2022)
Rice husk-derived	Mixed Cu^{2+}, Pb^{2+}, Cr^{3+}, Fe^{3+} solutions	Applicable	Vassileva et al. (2019)

(Coutinued)

TABLE 3.2
(Coutinued)

Adsorbent	Water or gaseous pollutant adsorbates	Applicability of Langmuir isotherm	References
Peanut shell	Pb^{2+}, Cd^{2+}, Ni^{2+}, Cu^{2+}, Zn^{2+}	Applicable	Shan et al. (2020)
Sugarcane bagasse biochar (SB)	Methylene blue (MB)	Applicable	Biswas et al. (2020b)
SB-alginate composite	MB, phenol	Applicable	Biswas et al. (2020b, 2020c)
Bamboo-based activated carbon	MB	Applicable	Yao et al. (2024)
Forest waste-based activated carbon	MB	Applicable	Yildiz (2024)
Java plum leaves	Malachite green (MG)	Applicable	Gul et al. (2024)
Bamboo-charcoal iron oxide Nanocomposite	MB	Applicable	Sen (2023a)
Biomass-based activated carbon (AC)	Phenolic compounds, heavy metal ions	Applicable	EI-Baz et al. (2020)
Carbon-based adsorbents, polymer-based adsorbents, and zeolite	CO_2, SO_2, NO_2, H_2S, VOCs	Applicable	Huang et al. (2019)
Biochar	CO_2, SO_2, Hg_2, VOCs, NH_3	Applicable	Zhao et al. (2022)
Magnetic iron oxide nanoparticles on Vasaka (*Justicia adhatoda*) leaf extract (JA-MIONs)	MB	Applicable	Ghosh et al. (2023)

Example 3.1: Adsorption isotherm experiment and applicability testing of equilibrium isotherm model

Adsorption isotherm experiments were conducted using a 40 mL reaction mixture with methylene blue (MB) dye concentrations ranging from 20, 30, 40, 50, 60 to 70 ppm, and 20 mg of adsorbent (pine tree leaves biomass powder). The experiments were carried out in a series of 250 mL conical flasks shaken in a constant temperature shaker for a period of three hours, which exceeded the equilibrium time, while all other physicochemical conditions were kept constant. After equilibrium was reached, the reaction mixture was centrifuged, and the residual MB dye concentrations (C_e) in mg/L were measured from the supernatant solution. The amount of dye adsorbed at equilibrium (q_e) in mg/g was determined using Eq. 3.2 (Sen, 2023b). The experimental adsorption equilibrium data obtained from this isotherm experiment are presented in Table 3.3. The separation factor (R_L) was calculated using Eq. 3.11. Test the applicability of Langmuir isotherm model and find out model parameters.

Solution: Using the data from Table 3.3, a linear plot of the Langmuir isotherm model equation, as per Eq. 3.9, was plotted between Ce/qe vs. Ce, which is presented in Figure 3.6. The obtained high linear regression coefficient (R^2) value of 0.9989 strongly indicates that the applicability of Langmuir isotherm model fits well.

From Figure 3.6, the slope of $1/q_m = 0.0271$; hence, the maximum adsorption q_m is 36.90 mg/g. From the intercept, the constant related to the binding energy of the sorption system (K) is 0.57 for the MB–pine tree leaves biomass system. The obtained dimensionless separation factor (R_L) using Eq. 3.11 at different initial dye concentrations was 0.128, 0.0685,

TABLE 3.3

Equilibrium adsorption data of methylene blue (MB) dye on pine tree leaves biomass (Sen, 2023b)

Equilibrium concentration, C_e (mg/L)	*Amount of adsorption at equilibrium, q_e (mg/g)*	*C_e (mg/L)*	*C_e/q_e*	*Separation factor, R_L*
2.1542	15.6916	2.1542	0.137284	0.128205
6.03295	27.9341	6.03295	0.215971	0.068493
14.321	31.358	14.321	0.456694	0.046729
23.8231	32.3538	23.8231	0.736331	0.035461
32.86523	34.26954	32.86523	0.959022	0.028571
42.8954	34.2092	42.8954	1.253914	0.023923
52.1576	35.6848	52.1576	1.46162	-

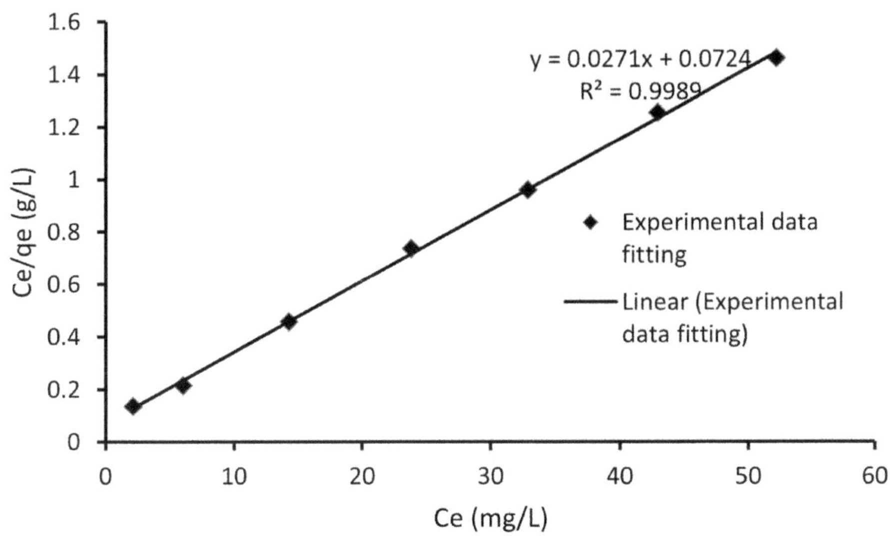

FIGURE 3.6

Linearized Langmuir isotherm model ditting plot with the adsorption equilibrium experimental data (solid/liquid adsorption system)

0.0467, 0.0355, 0.0286, and 0.0239, which confirmed the favourable adsorption of MB dye onto pine leaves biomass adsorbents, as all values lie in the range of 0 to 1 (Sen, 2023b).

Example 3.2: The following experimental data for the equilibrium adsorption of pure methane gas on activated carbon at 296 K are presented in Table 3.4. Test the applicability of the Langmuir adsorption isotherm for the given data (use a graph sheet). Find out the theoretical maximum adsorption capacity (q_m). The linearized Langmuir isotherm model is given as:

$$P/q = \frac{1}{q_m K} + \frac{P}{q_m}$$

Solution: The linearized form of Langmuir isotherm model Eq. 3.9 for gas/solid adsorption system can be written as

$$P/q = \frac{P}{q_m} + \frac{1}{K}$$

With help of above tabular adsorption equilibrium data (Table 3.4), a linearized plot of P/q vs. P (for gas/solid adsorption system) with high linear regression coefficient (R^2) value of 0.99 is shown in Figure 3.7.

From this plot intercept = 0.5682 = $1/(q_m{}^*K)$ and slope = 0.0073 = $1/q_m$.

Hence, the maximum adsorption capacity from slope q_m = 136.986 cm^3 is obtained.

Similarly, the Langmuir constant, K, from intercept = 0.5682 = $1/(q_m{}^*K)$

0.5682 = $1/(136.986^*K)$

$K = 0.012847$ $psia^{-1}$.

Example 3.3: The following experimental data for the equilibrium adsorption of pure CO onto charcoal at 1 atm pressure are presented in Table 3.5. Confirm that the following data fit a Langmuir isotherm. Find the constant K and the maximum adsorption capacity, q_m.

Solution: The linearized form of the Langmuir isotherm model (Eq. 3.9) for a gas/solid adsorption system can be written as:

$$P/q = \frac{1}{q_m K} + \frac{P}{q_m}$$

TABLE 3.4

Equilibrium adsorption data

q/cm^3	45.5	91.5	113	121	125	126	126
P, psia	40	165	350	545	760	910	970

J.D. Source: Seader (1998), Ernest J. Henley, Separation Process Principles, Page 800.

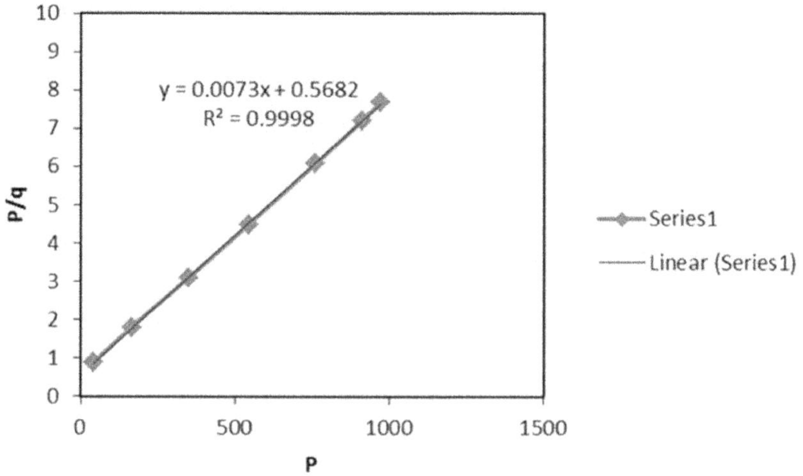

FIGURE 3.7
Linearized Langmuir isotherm model fitting plot with the adsorption equilibrium experimental data

TABLE 3.5
Equilibrium adsorption data of CO on charcoal

q/cm^3	10.2	18.6	25.5	31.5	36.9	41.6	46.1
P (torr)	100	200	300	400	500	600	700

TABLE 3.6
Equilibrium adsorption data of CO on charcoal

q/cm^3	10.2	18.6	25.5	31.5	36.9	41.6	46.1
P (torr)	100	200	300	400	500	600	700
P/q	9.8	10.8	11.8	12.7	13.6	14.4	15.2

Hence, using the Table 3.6 data, a plot of P/q against P should give a straight line of slope $1/q_m$ and intercept $1/Kq_m$, which is shown in Figure 3.8.

The plot yields a straight line, confirming that the data fit the Langmuir isotherm. The slope is 0.009 and hence maximum adsorption capacity, q_m = 1/0.009 = 111.11 cm^3. The intercept is 9.0143 and so K = 1/(111.11cm^3) × (9.0143 Torr cm^{-3}) = 9.98 × 10–^{4} $Torr^{-1}$.

3.2.1.3 Type II & III Freundlich Adsorption Isotherm

This is an empirical isotherm model known as the Freundlich isotherm model (1906). It shows large deviations from the Langmuir model. This isotherm

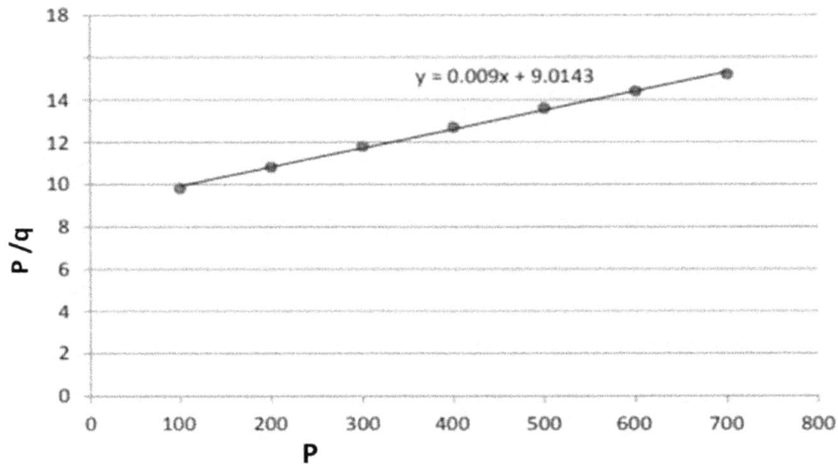

FIGURE 3.8
Linearized Langmuir isotherm model fitting plot with the adsorption equilibrium experimental data

model equation explains the formation of multilayer adsorption on a rough, heterogeneous surface, as depicted in Figure 3.9 (Dutta, 2007; Seader, 1996).

The Freundlich (1906) isotherm model assumes that the amount adsorbed at equilibrium (q_e), measured in mg/L for liquids or mg/surface area, has a power-law dependence on the partial pressure (or concentration) of the solute, as presented by Eq. 3.14 (Dutta, 2007):

$$q_e = K_F p^{\frac{1}{n}} \text{ for as/solid system} \tag{3.14}$$

The same equation can be written for a solid/liquid adsorption system as:

$$q_e = K_F \Big| Ce^{\frac{1}{n}} \tag{3.15}$$

From either Eq. 3.14 or Eq. 3.15, the linearized logarithmic model equations, Eqs. 3.16 and 3.17, can be written as follows:

$$\ln q_e = \ln K_F + \frac{1}{n} \ln P \tag{3.16}$$

or

$$\ln q_e = \ln K_F + \frac{1}{n} \ln C_e \tag{3.17}$$

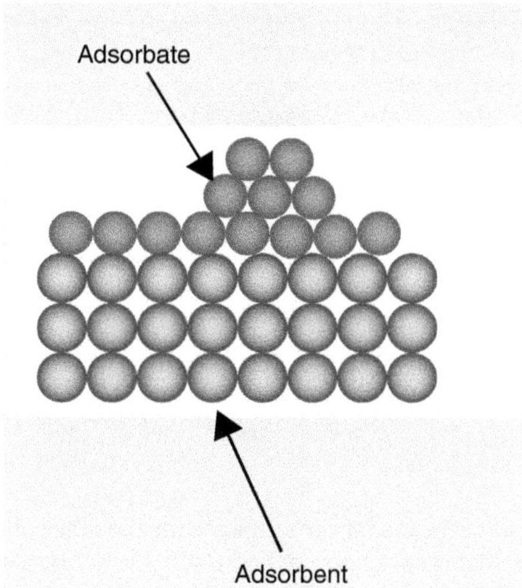

FIGURE 3.9
Multilayer adsorption

where q_e is the amount of dye adsorbed per unit mass of adsorbent at equilibrium (mg/g) and C_e is the equilibrium concentration of dye in solution (mg/L). 'K_F' and 'n' are the isotherm constants, which indicate the capacity and intensity of the adsorption, respectively (Dawood & Sen, 2012). The value of n typically varies between 1 and 5; for $n = 1$ the Freundlich model takes the form of Henry's Eq. 3.8 (Dutta, 2007). This is a two-parameter isotherm model, and the parameters 'K_F' and 'n' can be obtained from the intercept and slope of the linearized model fitting Eq. 3.16 or Eq. 3.17 with the isotherm experimental data set.

Dawood and Sen (2012) reported the Freundlich isotherm model fitting with their batch adsorption equilibrium study results for the pinecone–Congo red dye adsorption system at specific experimental conditions, as presented in Figure 3.10.

From Figure 3.10, the Freundlich isotherm model successfully explained the batch pinecone/CR dye equilibrium adsorption results, and it was applicable to the system, with a high linear regression coefficient (R^2) value of 0.95. The Freundlich constants, i.e., the adsorption capacity (K_F) and rate of adsorption (n), were calculated from the intercept and slope of the plot as 19.18 mg/g and 1.19, respectively. The value of n being greater than 1 indicates the favourable nature of the adsorption, and that the adsorption process is physical in nature (Dawood & Sen, 2012). Table 3.7 presents recent literature information on the applicability of the Freundlich isotherm model equation and its parameter values obtained from linear fittings with experimental data for different adsorbate-adsorbent systems,

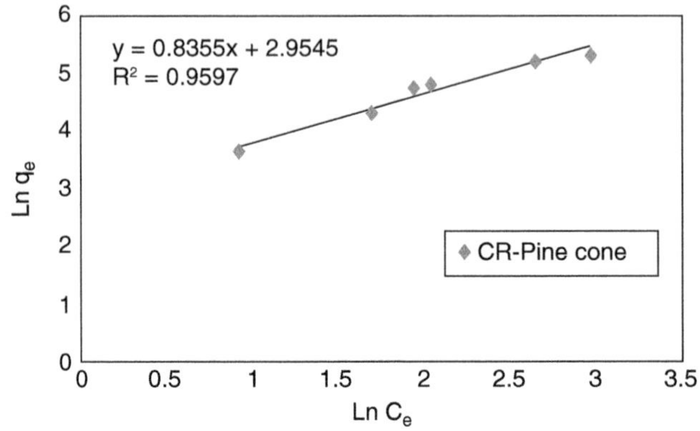

FIGURE 3.10
Linearized Freundlich isotherm model fitting plot with the adsorption equilibrium experimental data. Amount of adsorbent (pinecone) added = 10 mg; Initial CR dye concentration ranges: 10, 20, 30, 40, 50, 60 ppm; pH = 3.55; Temperature = 60°C; Shaker Speed = 120 rpm

Source: Taken from Dawood and Sen (2012) with written permission.

TABLE 3.7

Applicability of Freundlich isotherm model and its model parameters (n, K_F) values obtained from linear model fittings with high linear regression coefficient ($R^2 >$ 0.91) values for different solid/liquid adsorption systems at 25°C

Adsorbent	Adsorbate	n	K_F (mg/g)	Reference
Ca-alginate biochar	Zn^{2+}	1.25	3.27	Biswas et al. (2019a, 2019b)
Biochar-biopolymeric hybrid	Cu^{2+}	1.36	2.12	Biswas et al. (2019c)
Biochar-biopolymeric hybrid	Ni^{2+}	1.79	4.57	Biswas et al. (2019c)
Acid-activated pinecone biochar	Pb^{2+}	1.48	4.05	Biswas et al. (2020a)
Sugarcane bagasse biochar (SB)	Methylene blue (MB)	1.22	1.29	Biswas et al. (2020b)
SB-alginate composite	Phenol	4.61	5.20	Biswas et al. (2020c)
SB-alginate composite	Cd^{2+}	1.39	2.43	Biswas (2020d)
Polyethyleneimine-grafted graphene oxide	Pb^{2+}	3.65	19.07	Al-Yaari and Sahel (2023)
Bamboo-charcoal iron oxide nanocomposite	MB	1.30	3.205	Sen (2023a)
Polymer carbon nanofibre	Hg^{2+}	3.17	8.43	Al-Yaari et al. (2021)
Magnetic iron oxide nanoparticles on Vasaka (*Justicia adhatoda*) leaf extract (JA-MIONs)	MB	1.10	2.153	Ghosh et al. (2023)
JA-MIONs	Tetracycline hydrochloride (TCH)	1.18	1.429	Ghosh et al. (2023)

including single and mixed competitive adsorption systems, as reported by various researchers.

3.2.1.4 Extended Freundlich Isotherm Model

The extended Freundlich isotherm model is applicable for multicomponent multilayer adsorption on a solid adsorbent surface. The extended Freundlich isotherm models for two components in a binary system are given by Eqs. 3.18 and 3.19 (Bayuo et al., 2023; Papageorgiou et al., 2009):

$$q_{e,i} = \frac{K_{F,i} C_{e,i}^{(1/n_i)+x_i}}{C_{e,i}^{x_i} + y_i C_{e,j}^{z_i}} \tag{3.18}$$

and

$$q_{e,j} = \frac{K_{F,j} c_{e,j} \left(1/n_j\right) + x_j}{c_{e,i}^{x_i} + y_i c_{e,j}^{z_i}} \tag{3.19}$$

where $q_{e,\mathrm{i}}$ *and* $q_{e,j}$ are the amount of adsorption for components i and j (mg/g) at equilibrium, *Ce,i* and *Ce,j* are the equilibrium concentration of components *i* and j (mg/L). $K_{F,i}$ and $K_{F,j}$ are the Freundlich constants and n_i and n_j are the adsorption intensities obtained from the experimental data of the single component Freundlich isotherm models. The six parameters x_1, y_1, z_1 and x_2, y_2, z_2 are the multicomponent Freundlich model constants of two components.

3.2.1.5 Type II—Brunauer, Emmett, and Teller (BET) Isotherm Model

The BET isotherm model, which is a two-parameter model, explains the multilayer formation involved in physical adsorption. For a solid/liquid adsorption system, it can be represented either non-linearly or linearly, as given by Eqs. 3.20 and 3.21, respectively (Chen et al., 2022; Liu et al., 2024):

$$q_e = \frac{q_{mBET} C_{mBET} C_e}{(C_s - C_e)\left[1 + (C_{eBET} - 1)\left(\frac{C_s}{C_e}\right)\right]} \tag{3.20}$$

$$\frac{C_e}{(C_s - C_e) q_e} = \frac{1}{C_{BET} q_{mBET}} + \frac{C_{BET} - 1}{C_{BET} q_{mBET}} \left(\frac{C_e}{C_s}\right) \tag{3.21}$$

where q_{mBET} (mg/g) is the maximal sorption capacity and C_{BET} *(L/mg)* is the BET constant, which is linked toward the binding intensity for all layers. C_s (mg/L) is the saturation concentration. C_e is the equilibrium concentration.

For a gas/solid adsorption system, the non-linear BET isotherm model can be represented by Eq. 3.22. This type of adsorption isotherm for the gas/solid interface is classified as Type II.

$$V_{\text{total}} = \frac{V_{\text{mono}}\, C\left(\frac{P}{P_0}\right)}{\left(1-\frac{P}{P_0}\right)\left(1+C\left(\frac{P}{P_0}\right)-\frac{P}{P_0}\right)} \tag{3.22}$$

where V_{mono} = adsorbed volume of gas at high pressure conditions which helped to cover monolayer gas molecules formation on the solid surface and C = constant related to the energy of interaction with the surface.

Generally, various two-parameter and three-parameter adsorption isotherm model equations can be used separately or combined to evaluate the adsorption processes of various inorganics/organics at solid/liquid or gas/solid interfaces. The two-parameter and three-parameter isotherm model equations and their relevant model parameters are tabulated in Tables 3.8 and 3.9, respectively (Liu et al., 2024).

3.2.2 Surface Complexation Model (SCM) for Adsorption Mechanism

Since ion adsorption at the solid/liquid interface is generally more chemical than physical in nature, it is better to consider ion adsorption models based on chemical reactions or complexation models (Tien, 1994). The basis of the complexation model for ion adsorption is the formation of ion-surface complexes at charged sites on the solid adsorbent surface in solution. Surface Complexation Models (SCMs) have been shown to be capable of describing metal ion adsorption processes onto charged solid adsorbents, such as oxides and mineral surfaces (Crawford et al., 1996; Davis and Leckie, 1978; Dzombak & Morel, 1988, 1990; Katz and Hayes, 1995; Sen, 2001; Wen et al., 1998). These adsorption models assume that metal ions form complexes with surface functional groups in a manner similar to metal-ligand association in the solution phase. SCMs vary depending on the assumptions regarding the structure and composition of the sorbed species as well as on the physical description of the interfacial region. There are three main types of surface complex models, described below (Sen, 2001).

3.2.2.1 Constant Capacitance Model

The Constant Capacitance Model was suggested by Schindler and Stumm in Switzerland in the early 1970s (Hohl & Stumm, 1976; Schindler & Kamber, 1968; Sen, 2001). It is a special case of the diffuse double layer (DDL) model, developed for systems at high ionic strengths. In this model, the electrical

TABLE 3.8

2-Parameter adsorption isotherm model equation and their various model parameters

Model	Equation	Linear formula	Parameters	Graph
Langmuir isotherm	$q_e = \frac{Q_{mLan} b_{\text{Lan}} C_e}{1 + b_{\text{Lan}} C_e}$	$\frac{C_e}{q_e} = \frac{C_e}{Q_{mLan}} + \frac{1}{Q_{mLan} b_{\text{Lan}}}$	$Q_{mLan}\ (\text{mg/g})$, $b_{Lan}\ (\text{L/mg})$	$\frac{C_e}{q_e} \leftrightarrow C_e$
Extended Langmuir	$q_{e,i} = \frac{q_{m,EL} K_{EL,i} C_{e,i}}{1 + \sum_{j=1}^{N} K_{EL,i} C_{e,j}}$	$\frac{c_{e,i}}{q_{e,i}} = \frac{\sum_{j=1}^{N} K_{EL,i} C_{e,i}}{q_{m,EL} K_{EL,i}} + \frac{1}{q_{m,EL} K_{EL,i}}$	$q_{m,EL}, K_{EL,i}$	$\frac{C_{e,i}}{q_{e,i}} \leftrightarrow C_{e,i}$
Modified competitive Langmuir	$q_{e,i} = \frac{q_{m,i} K_{L,i} \left(\frac{c_{e,i}}{\eta_{L,i}}\right)}{1 + \sum_{N}^{j=1} K_{L,j} \left(\frac{c_{e,j}}{\eta_{L,j}}\right)}$	$\frac{\left(\frac{c_{e,i}}{\eta_{L,i}}\right)}{q_{e,i}} = \frac{\sum_{j=1}^{N} K_{EL,i} \left(\frac{c_{e,i}}{\eta_{L,i}}\right)}{q_{m,i} K_{L,j}} + \frac{1}{q_{m,i} K_{L,j}}$	$q_{m,i}, K_L$	$\frac{\left(\frac{c_{e,i}}{\eta_{L,i}}\right)}{q_{e,i}} \leftrightarrow \left(\frac{c_{e,i}}{\eta_{L,i}}\right)$
Freundlich isotherm	$q_e = k_{Fre} C_e^{\frac{1}{n}}$	$\ln q_e = \ln k_{Fre} + \frac{1}{n} \ln C_e$	$k_{Fre}\ [(\text{mg/g})\ (\text{mg/L})^{-1/n}], n$ x_i, y_i, z_i	$\ln k_{Fre} \leftrightarrow \ln C_e$
Extended Freundlich	$q_{e,i} = \frac{K_{EF,i} C_{e,i}^{(n_i + x_i)}}{C_{e,i}^{x_i} + y_i C_{e,j}^{z_i}}$ $q_{e,j} = \frac{K_{EF,j} c_{e,j}^{(n_j + x_j)}}{c_{e,j}^{x_j} + y_j c_{e,i}^{z_j}}$	-	x_i, y_i, z_i	

(Coutinued)

TABLE 3.8
(Coutinued)

BET isotherm	$q_e = \frac{q_{mBET} C_{mBET} C_e}{(C_s - C_e)\left[1 + (C_{eBET} - 1)\left(\frac{C_s}{C_e}\right)\right]}$	$\frac{C_e}{(C_s - C_e)q_e} = \frac{1}{C_{BET} q_{mBET}} + \frac{C_{BET} - 1}{C_{BET} q_{mBET}}\left(\frac{C_e}{C_s}\right)$	q_{mBET} , C_{BET}	$\frac{C_e}{(C_s - C_e)q_e} \leftrightarrow \left(\frac{C_e}{C_s}\right)$
D-R isotherm	$q_e = Q_{DR} \exp\left(-K_{DR}\left[RT\ln\left(1 + \frac{1}{C_e}\right)\right]^2\right)$	$\ln q_e = \ln Q_{DR} - K_{DR} R^2 T^2 \left[\ln\left(1 + \frac{1}{C_e}\right)\right]^2$	$Q_{DR}(mg/g)$, $K_{DR}(\text{mol}^2/\text{kJ}^2)$	$\ln q_e \leftrightarrow \left[\ln\left(1 + \frac{1}{C_e}\right)\right]^2$
Temkin isotherm	$q_e = \frac{RT}{b_{Tem}} \ln(K_{Tem} C_e) = B_{Tem} \ln(K_{Tem} C_e)$	$q_e = \frac{RT}{b_{Tem}} \ln(K_{Tem} C_e) = B_{Tem} \ln(K_{Tem} C_e)$	$b_{Tem}(\text{J/mol})$, K_{Tem}	$q_e \leftrightarrow \ln(C_e)$
H-J isotherm	$\left[\frac{1}{q_e^2}\right] = \left[\frac{B_{HJ}}{A_{HJ}}\right] - \left[\frac{1}{A_{HJ}}\right]\ln(C_e)$	$\left[\frac{1}{q_e^2}\right] = \left[\frac{B_{HJ}}{A_{HJ}}\right] - \left[\frac{1}{A_{HJ}}\right]\ln(C_e)$	B_{HJ} ,	$1/q_e{}^2 \leftrightarrow \ln(C_e)$
Halsey isotherm	$q_e = \text{Exp}\left[\frac{1}{n_{HL}}\left(\ln K_{HL} - \ln\left(\frac{1}{C_e}\right)\right)\right]$	$\ln(q_e) = \left[\left(\frac{1}{n_H}\right)\ln(K_H)\right] - \left(\frac{1}{n_H}\right)\ln\left(\frac{1}{c_e}\right)$	K_{HL}, n_{HL}	$\ln(q_e) \leftrightarrow \ln(C_e)$

Source: Taken from Liu et al., 2024 with written permission.

TABLE 3.9

3-parameter adsorption isotherm model equation and their various model parameters

Model	Equation	Parameters	Non-linear Graph
Sips isotherm	$q_e = \frac{K_S C_e^{\beta_S}}{1+\alpha_S C_e^{\beta_S}}$	K_S, α_S, β_S	$q_e \leftrightarrow C_e$
R-P isotherm	$q_e = \frac{K_{RP} C_e}{1+\alpha_{RP} C_e^{\beta}}$	$K_{RP}, \alpha_{RP}, \beta$	$q_e \leftrightarrow C_e$
Modified R-P	$q_e = \frac{K_{RP,i}\left(C_{e,i} / \eta_{RP,i}\right)}{1+\sum_{j=1}^{N} \alpha_{RP,j}\left(C_{e,j} / \eta_{RP,j}\right)^{\beta,j}}$	$K_{RP,i}, \alpha_{RP,j}, \beta_j$	$q_e \leftrightarrow C_e$
Toth isotherm	$q_e = \frac{q_T K_T C_e}{\left[1+(K_T C_e)^{nT}\right]^{1/nT}}$	q_T, K_T, n_T	$q_e \leftrightarrow C_e$
Radke-Prausnitz	$q_e = \frac{q_m K_{RP} C_e}{(1+K_{RP} C_e)^{m_{RP}}}$	K_{RP}, q_m, m_{RP}	$q_e \leftrightarrow C_e$
Fritz-Schluender	$q_e = \frac{q_m K_{FS} C_e}{1+q_m C_e^{m}}$	K_{FS}, q_m, m_{FS}	$q_e \leftrightarrow C_e$
Vieth and Sladek	$q_e = K_{VS} C_e + \frac{q_m \beta C_e}{1+\beta C_e} \times q$	K_{vs}, q_m, β	$q_e \leftrightarrow C_e$
Hill	$q_e = \frac{q_{HI} C_e^{nHI}}{K_{D,HI} + C_e^{nHI}}$	$K_{D,HI}, q_{HI}, n_{HI}$	$q_e \leftrightarrow C_e$

Source: Taken from Liu et al., 2024 with written permission.

double layer is treated as a parallel plate capacitor, i.e., it is based on the Helmholtz model. All adsorbing species are in one plane, meaning the same potential is experienced by all adsorbing ions (Cornell & Schwertmann, 1998; Wen et al., 1998).

3.2.2.2 Diffuse Double Layer Model

The DDL model is based on the Grouchy-Chapman model and was developed by Stumm et al. (1970). This model defines two planes in the interfacial region: a surface plane for adsorption of H^+, OH^-, and all specifically adsorbed species and a diffuse layer plane representing the closest distance of approach for all counterions.

The distribution of ions, charge, and potential in the diffuse layer is described by the Poisson-Boltzmann equation. The overall relationship between surface charge and potential is given by the Gouy-Chapman equation (Sen, 2001; Wen et al., 1998).

3.2.2.3 Triple Layer Model

The Triple Layer Model is similar to the extended Stern layer model, with the compact double layer split into two parts. In this model, there are three planes in the interfacial region: a surface plane for adsorption of H^+, OH^-, and strongly adsorbed ions; a near-surface plane (the β plane) for weakly adsorbed ions; and a diffuse layer plane, representing the closest distance of approach for dissociated charges (Sen, 2001).

Each of these models requires the determination of several parameters to describe both acid-base protolysis and metal ion sorption. Wen et al. (1998) successfully used SCMs to describe metal ion adsorption on pure mineral materials. In their study, the SCM approach was used to describe the surface properties and adsorption behaviour of natural aquatic sediment. They concluded that all three versions of the surface complex models—Constant Capacitance Model (CCM), Diffuse Double Layer Model (DDL), and Triple Layer Model (TLM)—could effectively interpret the experimental data for the pH adsorption edges for Cu and Cd metal ions. This suggests that all three SCMs are applicable for describing the adsorption behaviour of natural sediments. Katz and Hayes (1995) also successfully applied the TLM version of the SCM for Co^{2+} metal ion adsorption on Al_2O_3.

3.2.3 Development of a SCM for Metal Cation Adsorption (e.g., Ni^{2+} on Oxide Surface such as Iron Oxide, Fe_2O_3) (Sen, 2001; Sen et al., 2002; Sen, 2012)

A simple double-layer model based on the surface complexation mechanism has been developed to describe the experimental observations of Ni (II) adsorption on iron oxides (Sen et al., 2002; Sen, 2012). The oxide surface is assumed to have the following characteristics and properties:

- Each colloidal iron oxide particle can be represented by a single planar surface.
- The surface hydroxyl groups are uniformly distributed over the planar surface. These groups are amphoteric in nature.

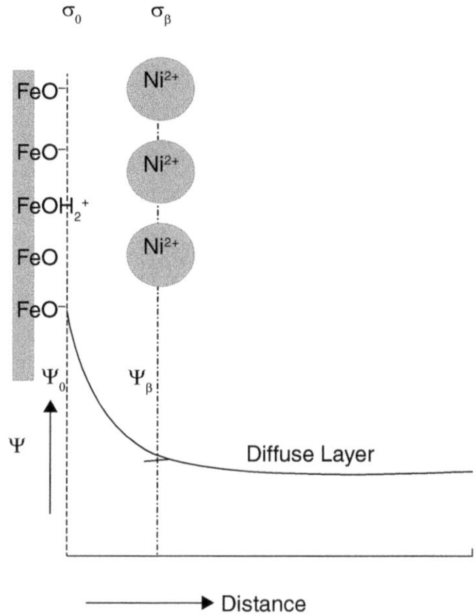

FIGURE 3.11
Schematic representation of the charge distribution at an idealized planar oxide surface and the potential decay away from the surface

Source: Taken from Sen, 2001, with written permission.

- Surface charge depends on pH, ionic strength, and the presence of specifically sorbing ions. When there is a net surface charge, an electrostatic potential develops, leading to ion distributions adjacent to the surface that compensate for the surface charge. This screening effect results in the electrostatic potential decaying with distance from the surface, forming an electric double layer. Such charge distributions are illustrated schematically in Figure 3.11.
- The equilibrium relationship for complexes formed between surface sites and solution species can be described in the same way as for complexes composed solely of solution species. However, the relevant activities of the species involved in surface reactions are those at the surface, not their activities in the bulk solution.

The surface charge on the iron oxide surface arises due to the dissociation (ionization) of surface hydroxyl groups. The following surface-changing mechanisms have been proposed based on the 2-pK model, as these functional groups are amphoteric in nature (Benjamin, 1981; Cornell and Schwertmann, 1998; Davis & Leckie, 1978; Mckinley & Jenne, 1991; Sen, 2001).

$$FeOH_2^+ \rightarrow FeOH^0 + H_s^+ \tag{3.23}$$

$$\mathrm{FeOH^0 \rightarrow FeO^- + H^+} \tag{3.24}$$

The equilibrium or dissociation constants are given below:

$$K_1 = \frac{\{\mathrm{FeOH^0}\}\{\mathrm{H_s^+}\}}{\{\mathrm{FeOH_2^+}\}} \tag{3.25}$$

$$K_2 = \frac{\{\mathrm{FeO^-}\}\{\mathrm{H_s^+}\}}{\{\mathrm{FeOH^0}\}} \tag{3.26}$$

In this model, we consider two planes at the interfacial region: a surface plane for the adsorption of H^+, OH^-, and all specifically adsorbed species and a diffuse layer plane representing the closest distance of approach for all counterions (Figure 3.11).

The magnitude and sign of the surface charge depend on the solution hydronium ion concentration and the dissociation constants K_1 and K_2. Since these dissociation reactions occur near a charged solid surface, the concentration of any ion is related to the bulk ion concentration by the Boltzmann equation:

$$\{\mathrm{H_x^+}\} = [\mathrm{H_b^+}]\exp\left(-\frac{e\psi_0}{k_B T}\right) \tag{3.27}$$

where $\{Hs^+\}$ = hydronium ion concentration near surface, $[H_b^+]$ = bulk hydronium ion concentration, Ψ_0 = surface potential at the plane, e = unit of electric charge, k_B = Boltzman constant, and T is absolute temperature.

Eqs. 3.25 and 3.26 can be written in terms of intrinsic equilibrium constant such as:

$$K_1^{\text{int}} = \frac{\{\mathrm{FeOH^0}\}[\mathrm{H_b^+}]}{\{\mathrm{FeOH_2^+}\}}\exp\left(-\frac{e\psi_0}{k_B T}\right) \tag{3.28}$$

$$K_2^{\text{int}} = \frac{\{\mathrm{FeO^-}\}[\mathrm{H_b^+}]}{\{\mathrm{FeOH^0}\}}\exp\left(-\frac{e\psi_0}{k_B T}\right) \tag{3.29}$$

{} and [] are surface and bulk phase concentrations, respectively. Adsorption of metal cation on oxide surface (say Ni^{2+}). A number of stoichiometric reactions could be written for adsorption of a metal cation on the iron-oxide surface. One such reaction is,

$$\mathrm{FeOH^0 + Ni^{2+} \rightarrow FeO - Ni^{2+} + H_s{}^+} \tag{3.30}$$

Eq. 3.30 can be written in terms of intrinsic adsorption constant:

$$K_{\text{ads}}^{\text{int}} = \frac{\left\{\text{FeO}^{-}\text{Ni}^{2+}\right\}\left[\text{H}_{\text{b}}^{+}\right]\exp\left(-\frac{e\psi_0}{k_B T}\right)}{\left\{\text{FeOH}^{0}\right\}\left[\text{Ni}^{2+}\right]\exp\left(\frac{e\psi_\beta}{k_B T}\right)} \tag{3.31}$$

For mathematical simplicity, we assume that the mean location of all specifically adsorbed ions is in the σ_β plane, and φ_β represents the mean electrostatic potential at that plane.

The total mole balance yields the following equation:

$$T_{\text{FeOH}} = F_t\left(\left\{\text{FeOH}^{0}\right\}+\left\{\text{FeO}^{-}\text{Ni}^{2+}\right\}+\left\{\text{FeOH}_2^{+}\right\}+\left\{\text{FeO}^{-}\right\}\right) \tag{3.32}$$

where F_t is the total moles of iron in suspension.

$$T_{\text{Ni}^{2+}} = \left[\text{Ni}^{2+}\right]_e + \left\{\text{FeO}^{-}\text{Ni}^{2+}\right\}\times F_t \tag{3.33}$$

The total number of sites (N_s) available for adsorption is given as:

$$N_s = \frac{N}{F_a}\left(\left\{\text{FeOH}_2^{+}\right\}+\left\{\text{FeO}^{-}\right\}+\left\{\text{FeOH}^{0}\right\}+\left\{\text{FeO}^{-}\text{Ni}^{2+}\right\}\right) \tag{3.34}$$

where N is the conversion factor from moles per liter to coulombs per square meter $\left(\frac{F_a V}{A_0 M}\right)$, F_a is the Faraday constant, M is the molecular weight, and A_0 is the BET surface area of the solid surface.

The surface charge σ_0 is given by:

$$\sigma_0 = N\left(\left\{\text{FeOH}_2^{+}\right\}-\left\{\text{FeO}^{-}\right\}-\left\{\text{FeO}^{-}\text{Ni}^{2+}\right\}\right) \tag{3.35}$$

$$\sigma_\beta = N\left(\left\{\text{FeO}^{-}\text{Ni}^{2+}\right\}\right) \tag{3.36}$$

Using a capacitor approximation, we can write:

$$\sigma_0 = C_1\left(\varphi_0 - \varphi_\beta\right) \tag{3.37}$$

The amount of adsorption (t) can therefore be obtained as:

$$\tau = \frac{T_{Ni} - \left[\text{Ni}^{2+}\right]_e}{F_t} = \left\{\text{FeO}^{-}\text{Ni}^{2+}\right\} \tag{3.38}$$

The adsorption maximum $(\tau_{\max})$ can be written as:

$$\tau^{\max} = \frac{T_{\text{FeOH}}}{F_t} = \left\{\text{FeOH}^0\right\} + \left\{\text{FeO}^-\text{Ni}^{+2}\right\} + \left\{\text{FeOH}_2\right\} + \left\{\text{FeO}^-\right\} \tag{3.39}$$

where $\tau_{\max}$ represents the fraction of F_t that could accept Ni^{2+} at saturation.

3.2.3.1 Comparisons of Model Predictions with Experimental Measurements

Sen (2001) found, both from model analysis and experimental results, that the adsorption of Ni (II) on iron oxide increases with solution pH, initial metal ion concentration, and loading rate. Therefore, for a particular pH and given equilibrium concentration of Ni^{2+}, the concentration of various species was calculated using a code developed in MATLAB from Eqs. 3.25, 3.26, 3.34, 3.37, and 3.38. In the model calculations, the values of σ_0, K_1, K_2, N_s, and C_1 were taken from literature sources (Benjamin & Leckie, 1981; Cornell & Schwertmann, 1998; Crawford et al., 1996; Davis et al., 1978), as presented in Table 3.10. The exact values of all these model variables are not known for our specific system. In the model, K_{ads} is the only adjustable parameter that is fitted with experimental values, as it is unknown.

TABLE 3.10

Base values are used in model calculations (Sen, 2001)

Parameter	**Value**
C_1	$140\ \mu F_a/cm^2$
N_s	11.4 sites $/nm^2$
T	298 K
ψ_0	−45 mV (pH = 11.5), 40 mV (pH = 4.5)
σ_0	$-5.24\times10^{-4}\,C/m^2$(pH = 11.5)
pK_1	7.2(pH = 11.5), 4.2(pH = 4.50)
pK_2	9.5
A_0	$191.6\ m^2/g$
F_a	96500 C mole
k_b	$1.38065\times10^{-23}\,J/K$
E	$1.602\times10^{-19}\,C$
Solid/Liquid	8 mg / 30 mL

Figure 3.12 shows the model predictions and the experimental isotherm for the Ni-iron oxide system at a pH of 11.5, with initial total Ni ion concentrations of 11.4, 12.4, 16.9, 20.4, and 24.5 ppm, and a solid/liquid ratio of 8 mg / 30 mL at room temperature. The model provides reasonable agreement with the experimental data reported by Sen et al. (2002), for which K_{ads} is equal to $10^{-7.8}$.

Figures 3.13 and 3.14 present the comparison between the model calculations and experimental data for the Ni-iron oxide system at a solution pH of 4.5, with a wide range of initial total metal ion concentrations from 10 to 800 ppm and a solid/liquid ratio of 8 mg / 30 mL at room temperature (Sen, 2001; Sen et al., 2002). Figure 3.14 presents the comparison between the isotherms for the same system at lower ranges of initial total Ni^{2+} concentrations, specifically 11.4, 12.4, 16.9, 20.4, and 24.5 ppm.

The simple surface complexation model is inconsistent with the observed maximum adsorption density, as reflected in the upper portion of Figure 3.13. The model predicts a continual increase in adsorption density even after the point of saturation. In addition to the deficiency of exact values for model parameters, experimental values indicate the presence of multiple types of sites with different sorption energies. This implies that the entire isotherm could be modelled section-wise for some portions using a Freundlich type and for others a Langmuir type. As a result, there will be different fittedK_{ads} values for different sections. Furthermore, there is

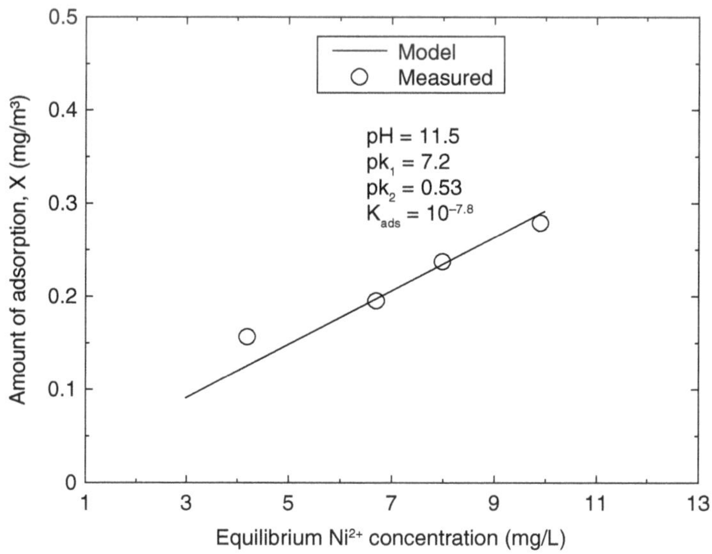

FIGURE 3.12
Comparison between model predictions and experimental data for the adsorption of Ni metal ion on iron oxide at pH 11.5 (Sen, 2001)

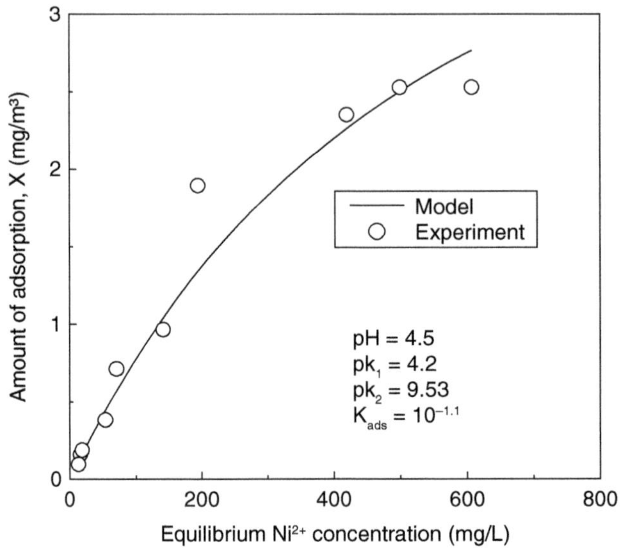

FIGURE 3.13
Comparison of model predictions with experimental data for the adsorption of Ni metal ion on iron oxide at pH 4.5 (Sen, 2001)

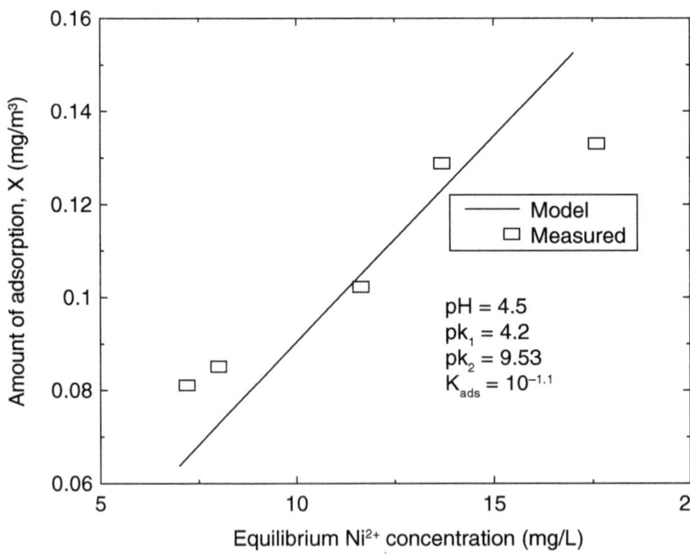

FIGURE 3.14
Comparison between model predictions and experimental data for the adsorption of Ni metal ion on iron oxide at pH 4.5 (lower metal concentration ranges)

Source: Sen, 2001.

also the possibility of surface complex formation, such as $FeO^{-}NiOH^{+}$. Therefore, the combination of these deficiencies should be incorporated into this simple surface complexation model to fully explain all experimental data.

3.3 Thermodynamics of Batch Adsorption and Parameter Estimation

Thermodynamic studies are used to investigate the effect of temperature on the adsorption process. The determination of the basic thermodynamic parameters—change in enthalpy of adsorption (ΔH^0), change in Gibbs free energy of adsorption (ΔG^0), and change in entropy of adsorption (ΔS^0)—is important for evaluating the feasibility of the process, assessing the spontaneous and favourable nature of the system, and ascertaining whether the process is exothermic or endothermic (Afroze & Sen, 2018; Yaneva and Georgieva, 2012). Numerous studies on adsorption have been published, involving adsorption isotherm and thermodynamic parameter analysis and calculations (Chen et al., 2021).

The change in Gibbs free energy of activation (ΔG^0), change in enthalpy of activation (ΔH^0), and change in entropy of activation (ΔS^0) are calculated using the following two well-known thermodynamic relations (Afroze & Sen, 2018; Tan & Sen, 2020):

$$\Delta G^0 = \Delta H^0 - T\Delta S^0 \tag{3.40}$$

$$\log\left(\frac{q_e}{c_\theta}\right) = \frac{\Delta S^0}{2.303\,R} + \frac{-\Delta H^0}{2.303\,RT} \tag{3.41}$$

where q_e is the solid phase concentration at equilibrium (mg/g), C_e is the equilibrium concentration in solution (mg/L), T is the temperature (K), and R is the gas constant (8.314 J/mol·K). Eq. 3.41 is known as the Van't Hoff equation. In Eq. 3.41, the equilibrium constant $K = (q_e / C_e)$, representing the adsorbate solute distribution between two different phases, has been reported with units of L/g by numerous published studies cited by Dawood and Sen (2012) and Tan and Sen (2020). This is due to the different solute phase concentration units: one in the solid adsorbent phase (q_e in mg/g) and the other in the liquid phase (C_e in mg/L) (Tan & Sen, 2020). However, the equilibrium constant $K = (q_e / C_e)$ must be recalculated as dimensionless by multiplying it by a factor of 1,000 (considering an adsorbent solution density of 1 g/mL), if it was originally expressed in L/g as reported by Milonjic (2007). Therefore, Eq. 3.41 should be rewritten as:

$$\log\left(1000\frac{q_e}{c_e}\right) = \frac{\Delta S^0}{2.303\,R} + \frac{-\Delta H^0}{2.303\,RT} \tag{3.42}$$

The entropy change (ΔS^0) and enthalpy change (ΔH^0) can be calculated from the intercept and slope of the linear Van't Hoff plot of $1000 \cdot \log\left(q_e / C_e\right)$ vs. $1/T$. From these values, the change in Gibbs free energy of activation (ΔG^0) can be calculated using Eq. 3.40 (Afroze & Sen, 2018; Tan & Sen, 2020).

There is considerable ambiguity in the calculation of adsorption equilibrium constants and thermodynamic parameters (ΔG^0, ΔH^0, and ΔS^0). Calculating these parameters requires a dimensionless standard equilibrium constant. The experimental equilibrium constant must be converted to the standard equilibrium constant according to the standard state. The choice of standard state affects the values of ΔG^0, ΔH^0, and ΔS^0. Readers are encouraged to refer to the article by Chen et al. (2021) for accurate calculations of various thermodynamic parameters from specific equilibrium adsorption model equations and parameters. The magnitude of the change in Gibbs free energy (ΔG^0) is commonly used to determine whether the adsorption process is controlled by physical adsorption (ranging from 0 to –20 kJ/mol) or by chemisorption (ranging from –400 to –80 kJ/mol) (Afroze & Sen, 2018; Cantuaria, de Almeida Neto et al., 2016; Zhau et al., 2012). A negative value of ΔH^0 indicates that the adsorption process is exothermic, while a positive value indicates an endothermic process (Afroze & Sen, 2018; Tan & Sen, 2019). A negative ΔS^0 indicates that the adsorbed molecules are in a more orderly state, and a negative ΔG^0 indicates that the adsorption process is spontaneous (Dutta, 2007; Tan & Sen, 2019).

Table 3.11 presents a compilation of recent studies on various thermodynamic parameter values for the adsorption of selected gaseous and water pollutants by various solid waste adsorbents. Zhou et al. (2012) reported thermodynamic parameters such as the change in Gibbs free energy (ΔG^0), change in enthalpy (ΔH^0), and change in entropy (ΔS^0), which were calculated to characterize the adsorption behaviour of SO_2, NO, and CO_2 in flue gas on activated carbon fibre (ACF) at temperatures ranging from 323.15 K to 363.15 K. Figure 3.15 shows that the change in free energy of the adsorption process at all temperatures was negative, indicating that the adsorption process was spontaneous, and that the free energy (ΔG) declines monotonically on ACF (Zhau et al., 2012). The order of values for the change in Gibbs free energy (ΔG) was as follows: $SO_2 > CO_2 > NO$, indicating that NO and CO_2 require less work to adsorb onto ACF than SO_2 (Zhau et al., 2012).

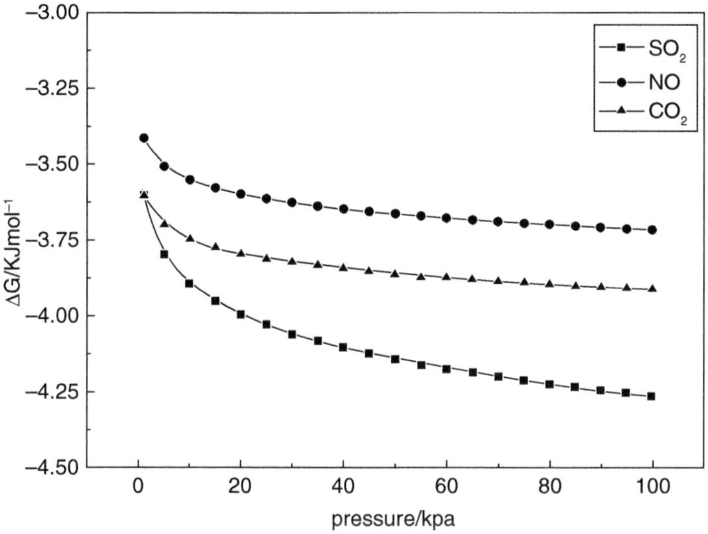

FIGURE 3.15
The change in Gibbs free energy (ΔG) for all adsorbates (SO_2, NO, and CO_2) on ACF

Source: Taken from Zhau et al., 2012, with written permission.

Similarly, Figure 3.16 presents the negative values of ΔS at all temperatures, indicating that the adsorption of SO_2, NO, and CO_2 proceeded from a random state to an ordered state on the surface of the adsorbent (ACF). Figure 3.16 shows that the ΔS values gradually decreased for SO_2 and NO over the entire pressure range, and the entropy of SO_2 and NO remained lower in negative values than that of CO_2 at all pressures (Zhau et al., 2012).

The adsorptions of SO_2, NO, and CO_2 onto ACF were exothermic in nature, which is indicated by the negative value of ΔH, as presented in Figure 3.17 (Zhau et al., 2012). The profiles of ΔH followed the same trend as $\Delta S \times T$. The ΔH values for SO_2 and NO decreased smoothly over all pressure ranges due to increased pressure (Zhau et al., 2012), whereas the ΔH values for CO_2 exhibited a smooth upwards trend at low equilibrium pressure (Figure 3.17). This is because the smaller molecular size of CO_2 favours delocalized adsorption, leading to efficient packing in the ACF cavity (Zhau et al., 2012). As pressure increased from 1 to 100 kPa, the ΔH values for SO_2 and NO ranged from –2 to –20 kJ/mol, indicating that physical adsorption and dipole-dipole interactions might be the main mechanisms of adsorption (Zhau et al., 2012).

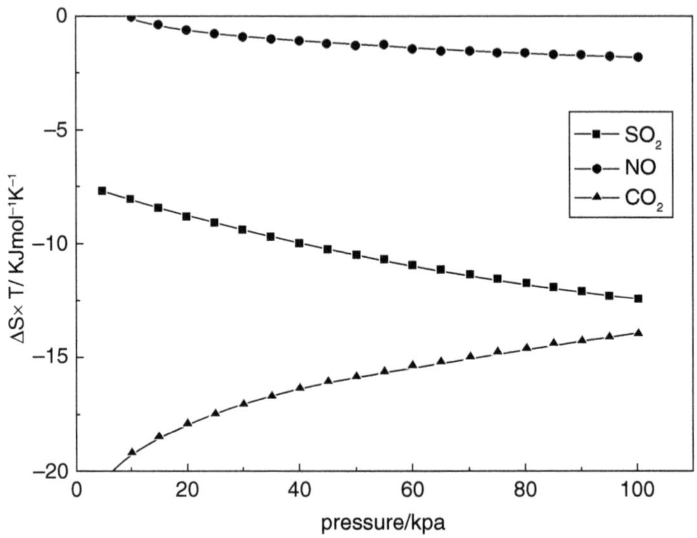

FIGURE 3.16
The change in entropy ($\Delta S \times T$) for all adsorbates (SO_2, NO, and CO_2) on ACF

Source: Taken from Zhau et al., 2012, with written permission.

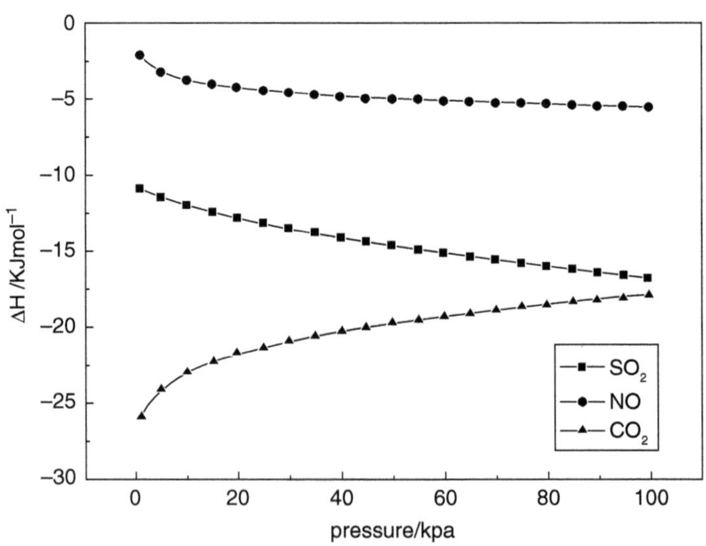

FIGURE 3.17
The change in enthalpy (ΔH) for all adsorbates (SO_2, NO, and CO_2) on ACF

Source: Taken from Zhau et al., 2012, with written permission.

TABLE 3.11

Thermodynamic parameter values for selected gaseous pollutants and water pollutants adsorption by various solid waste adsorbents recently reported by various researchers (modified from Afroze & Sen, 2018)

Adsorbents	Adsorbates (heavy metals and dyes)	Temerature range (K)	Thermodynamic parameters: Change in Gibbs energy, ΔG^0 (kJ/mole)	Change in enthalpy, ΔH^0 (kJ/mole)	Change in entropy, ΔS^0 (kJ/mole K)	Nature of adsorption	References
MWCNT/TiO_2 + Cr	SO_2	333–393	−6.67 to −4.07	−20.93	0.04	Exothermic	Zhang et al. (2020)
MWCNT/TiO_2 + Cu	SO_2	333–393	−2.86 to −1.79	−12.56	0.029–0.027	Exothermic	Zhang et al. (2020)
Coconut shell-based AC (SAC)	SO_2	323	−2.73	−2.61 to −10.6	−2.41 to −5.19	Exothermic	Honghong et al. (2014)
SAC	NO	323	−2.75	-	−0.26 to −18.15	-	Honghong et al. (2014)
CAC	CO_2	323	−2.76	-	−20.51 to −18.43	Exothermic	Honghong et al. (2014)
PEI/GO	SO_2	323	−2.75	−12.27 to −14.72	−9.79 to −8.95	Endothermic	Al-Yaari and Saleh (2023)
PCF (nanofibre)	Pb^{2+}	298–343	−1.00 to −4.15	19.73	0.07	Endothermic	Al-Yaari et al. (2021)
Magnetic activated carbon	Hg^{2+}	298–328	−0.66 to −4.56	38.07	0.13	Exothermic	Samanth et al. (2024)
Raw pomegranate peel	Herbicide	293–323	−5.37 to −5.02	−8.67	0.11	Endothermic	Ben-Ali, Jaouali et al. (2017)
	Cu^{2+}	303–313	-5.36 to -7.41	67.19	0.24		
Mango kernel (biocomposite)	Cr(VI)	306–334	−12.23 to −6.97	-68.34	-0.18	Exothermic	Akram, Bhatti et al. (2017)
Eucalyptus *sheathiana* bark (raw)	Zn^{2+}	303–333	−4.24 to −1.45	-32.43	-0.09	Exothermic	Afroze, Sen et al. (2016)

(Coutinued)

TABLE 3.11

(Coutinued)

Adsorbents	Adsorbates (heavy metals and dyes)	Temerature range (K)	Thermodynamic parameters			Nature of adsorption	References
			Change in Gibbs energy, ΔG^0 (kJ/mole)	Change in enthalpy, ΔH^0 (kJ/mole)	Change in entropy, ΔS^0 (kJ/mole K)		
Tephrosia purpurea leaf	Pb^{2+}	298–318	0.56 to 3.16	37.9	0.13	Endothermic	Madala et al. (2015)
Jackfruit leaf powder	Ni^{2+}	303–323	2.95 to −1.11	30.9	0.092	Endothermic	Boruah et al. (2015)
Sugarcane bagasse biochar (SB)	Methylene blue	303–333	−1.26 to −2.23	8.53	0.032	Endothermic	Biswas et al. (2020b)
Calcium alginate-SB	Methylene blue	303–333	−1.485 to −2.33	6.06	0.024	Endothermic	Biswas et al. (2020b)
Maize silk powder	Methylene blue	298–338	−1.62 to 0.25	-11.93	-0.03	Exothermic	Miraboutalebi et al. (2017)
Eucalyptus *sheathiana bark*	Methylene blue	303–333	−6.89 to −8.89	13.31	0.07	Endothermic	Afroze et al. (2016b)
Tomato seeds	Acid red 14	303–333	−6.45 to −5.09	−7.23	0.004	Exothermic	Najafi et al. (2015)
Peanut hull	Golden yellow dye	303–333	−2.10 to 1.73	−37.43	-0.12	Exothermic	Nawaz et al. (2014)
Pomelo (*Citrus grandis*) peel	Reactive blue 114	283–333	-42.51 to −69.88	89.47	0.48	Endothermic	Argun et al. (2014)
Pinecone-derived activated carbon	Congo red	303–333	−7.08 to −8.57	7.92	49.54	Endothermic	Dawood et al. (2014)

3.4 Kinetics of Batch Adsorption and Kinetic Parameters Determination

The design and scale-up of an adsorption process require a thorough understanding of adsorption kinetics and the adsorption mechanism. Generally, establishing adsorption kinetic models depends primarily on the amount of adsorption onto the surface of an adsorbent (q_t) as a function of operational time (t) (Liu et al., 2024; Tan & Hameed, 2017; Afroze & Sen, 2018). The nature of the adsorption processes such as solute uptake rate, rate control, and residence time must be evaluated under various physicochemical process parameters, including adsorbent dose, initial adsorbate concentration, ionic strength, solution pH, temperature profile, and contact time (Dawood, 2018).

The adsorption rate-controlling steps include external mass transfer, intraparticle mass transfer, solute diffusivity, and adsorptive reactions, which ultimately determine the best process conditions for full-scale operation (Afroze & Sen, 2018). Understanding the kinetics of dye and heavy metal ion adsorption onto adsorbent materials is a prerequisite for choosing the best operating conditions for a full-scale batch process (Yagub et al., 2014a). Various two-parameter kinetic models (2-PKMs) and three-parameter kinetic models (3-PKMs) of different complexity have been developed to predict the uptake rate (q_t) of solute adsorbate on the surface of a solid adsorbent (Liu et al., 2024; Tan & Hameed, 2017).

Adsorption kinetics can be represented by a plot of uptake (q_t) vs. time (t) under different physicochemical process parameters. For 2-PKMs, model parameters can be obtained based on the relationships of adsorption capacity (q_t) vs. operation time (t), i.e., $q_t \leftrightarrow t$, $q_t \leftrightarrow \sqrt{t}$, or $q_t \leftrightarrow \ln t$, using linear or non-linear fitting of curves (refer to Figure 3.18) (Liu et al., 2024). Various 2-PKM models, namely, Lagergren pseudo-first-order (PFO) and pseudo-second-order (PSO) kinetic models (Darweesh et al., 2022; Ho, 2006; Lagergren, 1898; Liu et al., 2024), the intraparticle diffusion (IPD) model (Azad et al., 2021; Liu et al., 2023; Liu et al., 2024), Ritchie modified second-order model (Behrooz et al., 2023; Liu et al., 2024; Zhang et al., 2023), Weber and Morris (WM) sorption kinetic model (Lima et al., 2022; Weber and Morris, 1963), Avrami fractional-order model (Liu et al., 2024; Vaghettie et al., 2009), Elovich equation (Barakakati et al., 2010; Zhang et al., 2019), and Bangham model (Liu et al., 2024), are presented in Figure 3.18. PFO and PSO models are the two most commonly used empirical models in liquid adsorption studies (Afroze & Sen, 2018; Sen, 2023; Tan & Hameed, 2017).

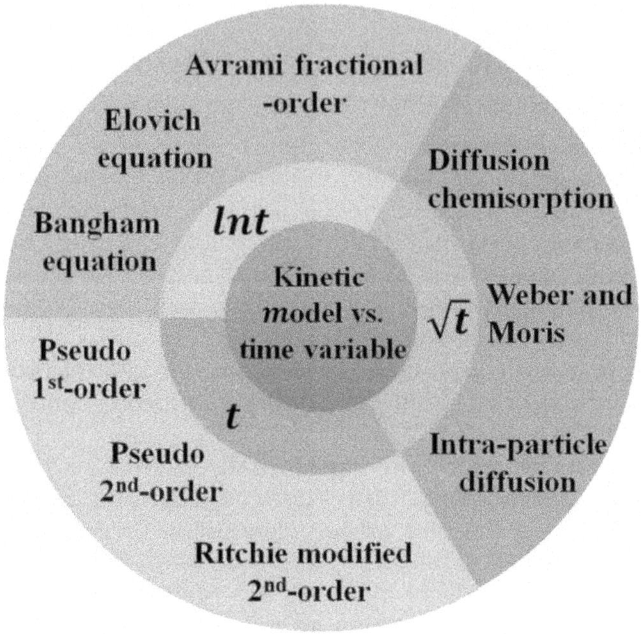

FIGURE 3.18
Various two-parameter sorption kinetic models vs. time variables

Source: Taken from Liu et al., 2024, with written permission.

3.4.1 Applicability of Various Kinetic Model Equations to Batch Kinetic Experimental Data Set and Determination of Kinetic Parameters and Error Function Analysis

Typically, a batch experiment is conducted to collect kinetic data under various physicochemical process conditions. Ensuring constant experimental conditions during batch adsorption is important for fitting model equations and estimating parameters. To avoid the iterative model parameter estimation procedure, a simple, common, but traditional practice in environmental adsorption studies is to fit the linearized form of various nonlinear kinetic model equations to experimental data sets using traditional linear least squares techniques with acceptable minimum error estimation (Tan & Hameed, 2017; Tan & Sen, 2020). The applicability of the best-fitted model equation is determined based on the higher linear regression coefficient (R^2) value, and the obtained model parameters are cross-checked with experimental values. Fitted kinetic parameters are determined by regression to minimize the selected error functions between predicted and experimental values (Tan & Hameed, 2017; Tan & Sen, 2020). However, non-linear regression of kinetic model equations with batch experimental data is recommended for accurate determination of different kinetic model parameters. It has been shown to be superior to linear regression, resulting in more realistic values of q_e and k (Tan & Hameed, 2017). Tan and Hameed (2017) critically discussed and clarified the indiscriminate applications, ambiguities, and controversies surrounding linearized kinetic model analysis vs. non-linear model analysis in their review article. They also discussed the appropriateness of linear regression for correlating kinetic data (Tan & Hameed, 2017).

With the advent of computing technology, non-linear regression analysis of various nonlinear model equations is highly recommended over linear regression. Linear regression analysis is a traditional and commonly used approach, allowing kinetic and isotherm model parameter values to be obtained quickly and easily, with an acceptable error range, for small batch solid/liquid adsorption systems (Tan & Sen, 2020).

Error functions are statistics that quantify the error between calculated model parameters and experimental values. Linear regression is a standard method developed based on the least squares criterion (Montgomery et al., 2012). Generally, during linear model applicability, these error statistics are calculated in addition to determining the linear regression coefficient (R^2) to confirm and support the model discriminated by R^2. Higher R^2 values generally correspond to lower error statistics (Tan & Hameed, 2017). Some commonly used error functions during model applicability determination are as follows (Tan & Hameed, 2017).

- Sum of squares of error (SSE)

$$\mathrm{SSE} = \sum_{i=1}^{n} (q_{\mathrm{cal}} - q_{\mathrm{exp}})^2 \tag{3.43}$$

- Chi-squared statics, χ^2

$$\chi^2 = \sum_{i=1}^{n} \frac{(q_{\mathrm{exp}} - q_{\mathrm{cal}})^2}{q_{\mathrm{cal}}} \tag{3.44}$$

- Mean sum-of-squares error (MSE)

$$\mathrm{MSE} = \sum_{i=1}^{n} (q_{\mathrm{cal}} - q_{\mathrm{exp}})^2 \tag{3.45}$$

- Root mean sum-of-squares error (RMSE)

$$\mathrm{RMSE} = \sqrt{\frac{1}{n} \sum_{i=1}^{n} (q_{\mathrm{exp}} - q_{\mathrm{cal}})^2} \tag{3.46}$$

where q_{cal} = amount of equilibrium adsorption from model predicted value and q_{exp} = experimentally obtained amount of equilibrium adsorption. Subscript i is omitted from q_{cal} and q_{exp} for clarity.

3.4.1.1 Lagergren Pseudo-First Order (PFO) and Pseudo-Second Order (PSO) Adsorption Kinetic Models

Lagergren (1898) derived the PFO equation to calculate the rate of adsorption, as presented by the following differential equation (Eq. 3.47) (El-Baz et al., 2020; Liu et al., 2024; Sen, 2023; Tan & Hameed, 2017):

$$\frac{dq_t}{dt} = k(q_e - q_t) \tag{3.47}$$

where q_t is the adsorption capacity (mg/g) at time t; t is time (min); q_e is the adsorption capacity at equilibrium; and k is the rate constant (1/min).

By integrating Eq. 3.47 at boundary conditions (BCs) when $t = 0$, $q_t = 0$ and at $t = t(> 0)$, $q_t = q_{t^-}$ the linearized form is obtained as:

$$\log(q_e - q_t) = \log q_e - \frac{k}{2.303}t \tag{3.48}$$

The slope and intercept of a linear plot of $\log(q_e - q_t)$ vs. t give the value of k and q_e, respectively. A straight line of $\log(q_e - q_t)$ vs. t with a high value of linear regression coefficient (R^2) indicates the applicability of this model equation to the experimental values under specified conditions.

Al Yaari et al. (2023) synthesized polyethyleneimine-grafted graphene oxide (PEI/GO) nanomaterials and tested their effectiveness in removing aqueous phase Pb^{2+} metal ions by adsorption at different initial metal ion concentrations, solution pH, dose, and temperature. They reported the adsorption

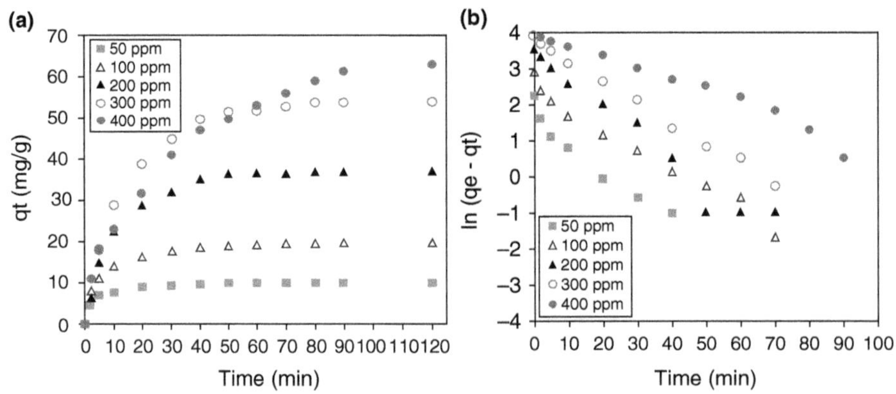

FIGURE 3.19

Pb^{2+} Adsorption kinetic studies at different initial metal ion concentrations using PEI/GO adsorbents (a) experimental data, (b) linear pseudo-first-order (PFO) model fitting. Solution pH = 6, dose = 0.1 g, temperature = 23°C

Source: Taken from Al-Yaari et al. (2023) with written permission.

TABLE 3.12

Fitted pseudo-first order (PFO) kinetic model parameters for adsorption of Pb^{2+} metal ions by PEI/GO at different initial metal ion concentrations (Al-Yaari et al., 2023)

Initial Pb^{2+} Metal Ion Concentrations, C_0(mg / L)	q_e, Exp (mg / g)	q_e cal (mg / g)	k (min^{-1})	R^2
50	9.98	9.70	0.246	0.9391
100	19.8	18.94	0.174	0.9751
200	37	36.53	0.091	0.9625
300	53.6	52.80	0.074	0.9951
400	62.8	60.81	0.040	0.9748

kinetic results at different initial metal ion concentrations, keeping other parameters constant. It was found that the linear PFO model (Eq. 3.48) fitted well with the experimental data, with a high linear regression coefficient (R^2) value over 0.98, as shown in Figure 3.19. The slope and intercept were determined from the linear fitted PFO model plot of $\ln(q_e - q_t)$ vs. time (t) (Figure 3.19), providing values for k_1 and q_e (calculated). The various fitted PFO kinetic parameters obtained from Figure 3.19 are tabulated in Table 3.12.

From Table 3.12, the high values of regression coefficients (R^2) in the range of 0.93 to 0.97 indicate the applicability of the PFO model to the batch adsorption kinetic experimental data. Furthermore, it can be observed that the calculated amount of Pb^{2+} adsorption at equilibrium (q_e) from both linear fitting data and the experimental amount of adsorption at equilibrium (q_e) are very close to each other. This very low error function further confirms the applicability of the PFO model for this adsorption system.

> **Example 3.4:** Batch adsorption kinetic experiments on aqueous phase inorganic Ni^{2+} metal ion removal by pinecone biochar adsorbent were conducted at different initial adsorbate metal ion concentrations. A fixed amount of 20 mg adsorbent was mixed with 50 mL(0.05 L) of adsorbate of predetermined concentration in a series of 125 mL labelled plastic bottles. The mixture was shaken in a constant temperature shaker using a Thermos Line Scientific Orbital Shaker Incubator at a speed of 130 rpm and a temperature of 35°C. The other process parameters were kept constant, such as solution pH = 9.4. At predetermined time intervals, the suspensions were taken out, and the remaining adsorbate concentration was measured using the Shimadzu AA-7000 Atomic Absorption Spectrophotometer (AAS) for Ni^{2+} ion solutions. The amount of adsorbate adsorbed onto the adsorbent at time t, q_t(mg / g), and adsorption

percentage were calculated using Eqs. 3.2 and 3.3, respectively. In this experiment, Ni^{2+} ion concentrations were varied from 10 to $70\ mg/L$. The following kinetic experimental data on metal ion adsorption at different initial adsorbate concentrations are tabulated in Tables 3.13–3.16, respectively.

a. Explain the effect of initial metal ion concentration on adsorption kinetics.
b. Test the applicability of pseudo-first order (PFO) kinetic model equation to the experimental kinetic data and determine the various kinetic model parameters (source-taken from Dawood, 2018; Dawood et al., 2017).

Solution:

Calculation of amount of adsorption, q_t (mg/g) and % of adsorption at each time, *t*.

Sample calculation, at time, $t = 5$ min, by using Eqs. 3.2 and 3.3

$$\text{Amount of adsorption, } q_t(\text{mg/g}) = \frac{(C_0 - C_e)V}{m} = (10 - 2.64)0.05/0.02 = 18.4$$

$$\%\text{ removal of adsorption, } = \frac{C_0 - C_e}{C_0} \times 100 = 73.6$$

Now new Table 3.14 is generated with the additional two columns, one for q_t and another for % of adsorption are as follows:

The applicability of PFO model, Eq. 3.48, was tested to the experimental Ni^{2+} adsorption by pinecone biochar data set, as presented in Table 3.14. A fitted straight line of log ($q_e - q_t$) vs. *t* with the batch kinetic experimental data at different initial Ni^{2+} metal ion concentrations (C_0) is presented in Figure 3.20.

TABLE 3.13

Kinetic Ni^{2+} adsorption data on pinecone biochar adsorbent (Dawood, 2018) Solution pH = 9.4, adsorbent dose = 20 mg, reaction mixture volume = 50 ml, temp. =35°C

Time, *t* (min)	Remaining Ni^{2+} metal ion concentration, C_t (mg/L) at time, t	Initial adsorbate Ni^{2+} metal ion concentration, C_0 (mg/L)
0	0.0	10
5	2.64	
15	2.23	
30	2.18	
45	2.17	
60	2.14	
90	2.02	
120	1.96	

(Coutinued)

TABLE 3.13

(Coutinued)

Time, *t* (min)	Remaining Ni^{2+} metal ion concentration, C_t (mg/L) at time, t	Initial adsorbate Ni^{2+} metal ion concentration, C_0 (mg/L)
0	0.0	20
5	6.83	
15	6.12	
30	6.08	
45	5.98	
60	5.45	
90	5.01	
120	4.99	
0	0.0	30
5	10.21	
15	8.57	
30	8.52	
45	8.23	
60	8.06	
90	8.02	
120	7.92	
0	0.00	50
5	24.25	
15	21.44	
30	20.04	
45	19.57	
60	19.38	
90	19.18	
120	19.16	
0	0.00	70
5	40.64	
15	37.10	
30	35.39	
45	34.67	
60	34.21	
90	34.17	
120	34.15	

The slope and intercept were determined from the linear fitted PFO model plot of log ($q_e - q_t$) vs. t (Figure 3.20), which will give the value of k_1 and q_e (cal), respectively. The error function chi-square (χ^2) was determined from the q_e (cal) verses q_e (exp) as per Eq. 3.44. Therefore, the various fitted parameters from Figure 3.20 are tabulated in Table 3.15. The lower values of linear regression coefficients (R^2) and high values of chi-square (χ^2) error function from Table 3.15 at various initial metal ion concentrations indicated the inapplicability of this PFO kinetic model. Additionally, the PFO model predicted a significantly lower amount of metal ion adsorbed at equilibrium, qe (mg/g), from the experimental value.

TABLE 3.14

Kinetic Ni^{2+} adsorption data on pinecone biochar adsorbent (Dawood, 2018)
Solution pH = 9.4, adsorbent dose = 20 mg, reaction mixture volume = 50 ml, temp. = 35°C

Time, t (min)	*Remaining Ni^{2+} metal ion concentration, C_t (mg/L) at time, t*	*Amount of adsorption, q_t (mg/g)*	*% of adsorption*	*Initial adsorbate Ni^{2+} metal ion concentration, C_0 (mg/L)*
0	0.0	0.00	0.0	10
5	2.64	18.40	73.60	
15	2.23	19.42	77.70	
30	2.18	19.54	78.60	
45	2.17	19.58	78.31	
60	2.14	19.64	78.56	
90	2.02	19.96	79.83	
120	1.96	20.11	80.43	
0	0.0	0.00	0.00	20
5	6.83	32.93	65.86	
15	6.12	34.71	69.42	
30	6.08	34.80	69.61	
45	5.98	35.06	70.11	
60	5.45	36.39	72.77	
90	5.01	37.47	74.94	
120	4.99	37.52	75.04	
0	0.0	0.00	0.00	30
5	10.21	49.48	65.97	
15	8.57	53.58	71.44	
30	8.52	53.69	71.59	
45	8.23	54.42	72.57	
60	8.06	54.85	73.13	
90	8.02	54.94	73.26	
120	7.92	55.21	73.61	
0	0.00	0.00	0.00	50
5	24.25	64.37	51.50	
15	21.44	71.40	57.12	
30	20.04	74.91	59.93	
45	19.57	76.08	60.86	
60	19.38	76.54	61.24	
90	19.18	77.06	61.65	
120	19.16	77.11	61.69	
0	0.00	0.00	0.00	70
5	40.64	73.41	41.95	
15	37.10	82.41	47.00	
30	35.39	86.52	49.44	
45	34.67	88.32	50.47	
60	34.21	89.47	51.12	
90	34.17	89.56	51.18	
120	34.15	89.63	51.22	

Table 3.15

The various fitted pseudo-first-order (PFO) kinetic model parameters for adsorption of Ni^{2+} metal ions by pinecone biochar at different initial metal ion concentrations

Initial Ni^{2+} metal ion concentrations, C_0 (mg/L)	$q_{e,\,Exp}$ (mg/g)	$q_{e,\,cal}$ (mg/g)	k (min^{-1})	R^2	Chi-Square, χ^2
10	20.11	2.03	0.05	0.816	161.19
20	37.52	3.81	0.07	0.902	298.17
30	55.21	2.14	0.03	0.474	1318.83
50	77.11	1.75	0.06	0.739	3250.09
70	89.63	1.41	0.03	0.243	5536.20

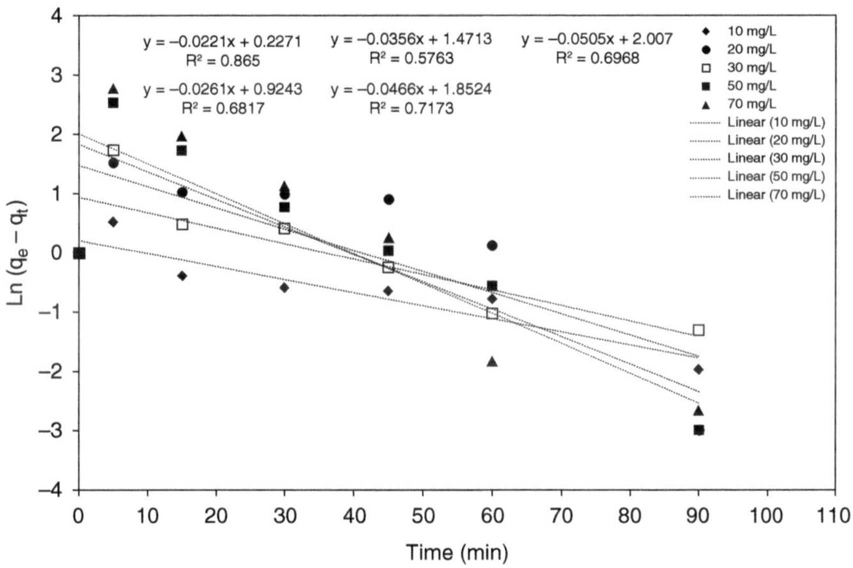

FIGURE 3.20
Linear fitted pseudo-first-order kinetic model for the adsorption of Ni (II) ions by pinecone biochar at a) different initial Ni^{2+} metal ion concentrations

Source: Taken from Dawood, 2018.

3.4.1.2 PSO Adsorption Kinetic Model

This model assumes that the uptake rate is second order with respect to the available surface sites (Ho & McKay 1999; Tan & Hameed, 2017), which is presented in Eq. 3.49.

$$\frac{dq}{dt} = k_2\left(q_e - q\right)^2 \tag{3.49}$$

Integrating Eq. 3.49 with initial conditions of $q = 0$ when $t = 0$ and subsequent rearrangement obtains the linearized form as per Eq. 3.50 (Afroze & Sen, 2018; Tan & Hameed, 2017).

$$\frac{t}{q_t} = \frac{1}{k_2 q_e^2} + \frac{1}{q_e} t \tag{3.50}$$

where k_2 (g mg^{-1} min^{-1}) is PSO rate constant of adsorption. A plot between t/q_t vs. t gives the value of rate constant k_2 (g/mg min), initial sorption rate h (mg/g-min), and theoretical q_e (mg/g) can be calculated. A plot of t/q_t against t should give a linear relationship for the applicability of the second-order kinetic (Afroze & Sen, 2018). The slope is $1/q_e$, and the intercept is $1/k_2\ q_e^2$. Other linearized forms are available but less frequently used (Ho, 2006).

The constant k_2 is used to calculate the initial sorption rate h, at $t \rightarrow 0$, as follows:

$$h = k_2 q_e^2 \tag{3.51}$$

Thus, the rate constant k_2, initial adsorption rate h, and predicted q_e can be calculated from the plot of t/q_t vs. time t using Eq. 3.50.

Most of the batch adsorption kinetic experimental data set under various physiochemical process conditions are fitted well by PSO, thereby indicating its superiority to other models (Afroze & Sen, 2018; Sen, 2023a; Tan & Hameed, 2017). The constant k_2 is a time scale factor that decreases with increasing initial adsorbate concentration (C_0). If the PSO model equation is fitted well to experimental kinetic data, the adsorption process becomes chemical in nature, whereas if PFO model is applicable then the adsorption process is physical in nature (Liu et al., 2024).

Biswas et al. (2020a) tested the applicability of PSO kinetic model to their experimental batch Pb^{2+} adsorption data on acid-treated pinecone biochar composite adsorbent system at various process parameters and reported the comparative results obtained from both non-linear and linear Eqs. 3.49 and 3.50 fitting data which is presented in Figure 3.21. The various fitted PSO kinetic model parameters for adsorption of Pb^{2+} metal ions by acid-treated pinecone biochar at different initial metal ion concentrations are tabulated in Table 3.16. The very high values of regression coefficients and very low values of error function indicate the applicability of PSO model from Table 3.16. Furthermore, it can be noticed from Table 3.16 that the calculated amount of Pb^{2+} adsorption at equilibrium (q_e) from both linear and non-linear model fitting data and the experimental amount of adsorption at equilibrium (q_e) are very close to each other, and hence, very low error function gives the additional confirmation for the applicability of this PSO model. Interestingly there is no significant differences in values between linear and non-linear fitting model methods and this is because of experimental run with lower initial metal ion concentration ranges.

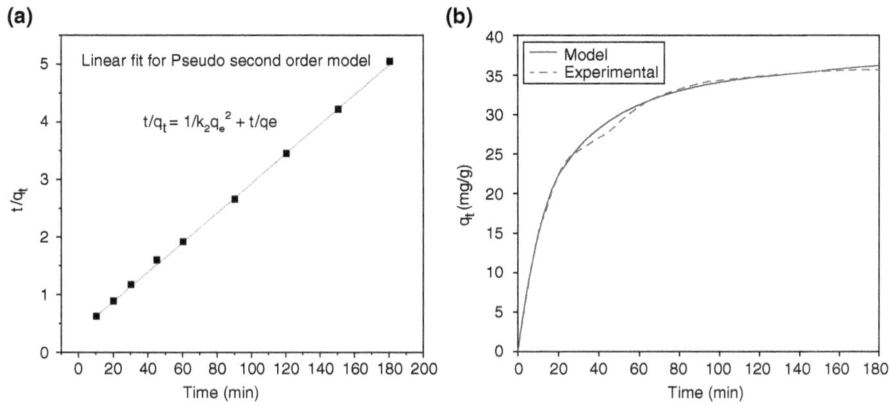

FIGURE 3.21
Typical pseudo-second-order kinetic model for a) liner fit model and b) non-linear fit model

Source: Biswas et al., 2020d.

> **Example 3.5:** Test the applicability of PSO model to the same batch adsorption kinetic data presented in **Example 3.4** and determine the various kinetic parameters.

Solution: PSO kinetics model was applied to investigate the applicability of Ni^{2+} ion adsorption onto pinecone biochar as per Eq. 3.50. A plot of t/q vs. t gives a straight line for PSO-compliant kinetics, which is presented in Figure 3.22 for the same Ni^{2+} adsorption kinetics data onto pinecone biochar at different initial metal ion concentrations and at different doses.

From Figure 3.22, it was found that the experimental data were fitted well into the PSO kinetic model with very high linear regression coefficients (R^2) values of greater than 99% at different metal ion concentrations and doses respectively. Also, very low chi-square (χ^2) values were calculated lower than 0.004 for various process parameters, as shown in Table 3.17. From Figure 3.22, the amount of adsorption at equilibrium (q_e), PSO rate constant (k_2), value of linear regression coefficient (R^2), chi-square, χ^2, and h values are tabulated in Table 3.17. From Table 3.17, the amount of Ni^{2+} metal adsorption using pinecone biochar adsorbent increased with the increase in metal ion concentration, but the rate constant, k_2, decreased with the increase in metal ion concentration. The initial adsorption rate h (mg/g.min) was calculated as per Eq. 3.51 and it was observed that the initial rate of adsorption was increased with the adsorbate metal ion concentration. Furthermore, it can be noticed from Table 3.17 that the calculated amount of Ni^{2+} adsorption at equilibrium (q_e) from model fitting data and the experimental amount of adsorption at equilibrium (q_e) are very close to each other, and hence, very low error function gives the additional confirmation for the applicability of this PSO model.

TABLE 3.16

The various fitted pseudo-second order (PSO) kinetic model parameters for adsorption of Pb^{2+} metal ions by acid-treated pinecone biochar at different initial metal ion concentrations

Adsorbate	$q_{e,Exp}$ *(mg/g). Experimental value*	q_e *(mg/g), calculated from linear fitting*	$q_{e,cal}$ *(mg/g), calculated from non-linear fitting*	k_2 *(g/mg-min). from linear fitting*	k_2 *(g/mg.min) from non-linear fitting*	R^2	*Chi-square,* χ^2	*h (mg/g.min)*
Pb^{2+}	11.95	11.25	12.02	1.59×10^{-3}	1.88×10^{-3}	0.99	0.043	0.297
	22.16	23.60	22.24	1.43×10^{-3}	1.78×10^{-3}	0.99	0.093	0.865
	30.9	32.1	31.04	1.37×10^{-3}	1.21×10^{-3}	0.99	0.039	1.5
	38.42	39.02	38.5	5.54×10^{-4}	5.74×10^{-4}	0.99	0.61	1.72

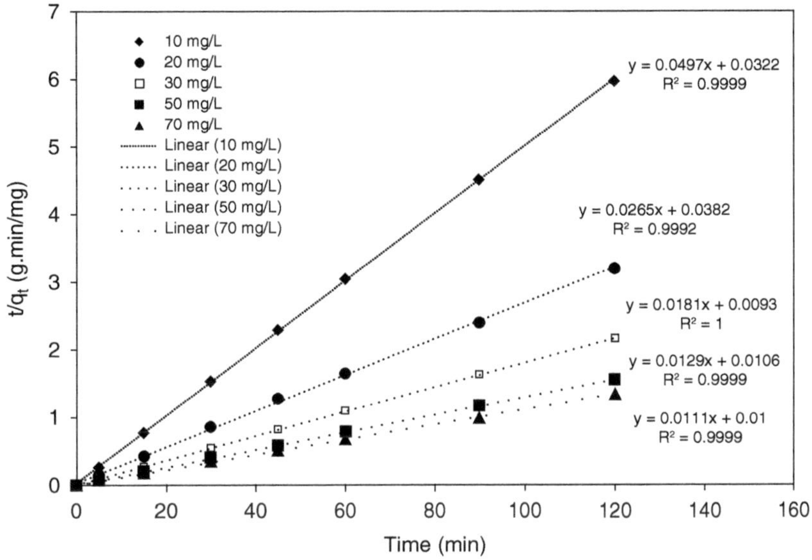

FIGURE 3.22
Linear fitted pseudo-second-order (PSO) kinetic model for the adsorption of Ni (II) ions by pinecone biochar at (a) at different adsorbent doses and (b) at different initial Ni^{2+} metal ion concentrations

Source: Taken from Dawood, 2018, Dawood et al., 2017.

TABLE 3.17
The various fitted pseudo-second-order (PSO) kinetic model parameters for adsorption of Ni^{2+} metal ions by pinecone biochar at different initial metal ion concentrations

Initial Ni^{2+} metal ion concentrations, C_0 (mg/L)	$q_{e,Exp}$ (mg/g)	$q_{e,cal}$ (mg/g)	k_2 g/ mg-min	R^2	Chi-square, χ^2	h (mg/g. min)
10	20.11	20.12	0.077	0.999	0.00	31.25
20	37.52	37.74	0.018	0.999	0.001	26.18
30	55.21	55.25	0.035	0.999	0.000	107.53
50	77.11	76.92	0.188	0.999	0.000	1111.11
70	89.63	89.29	0.139	0.999	0.001	1111.11

Example 3.6: Francisco did his final year research project batch experiment on Zn (II) adsorption on kaolin at a solution pH of 5.8. The reaction volume was 0.025 L and mass of adsorbent was 0.1 gram. He was reported the following kinetic experimental data on metal ion adsorption at two different initial metal ion concentrations, which is presented in Table 3.18.

TABLE 3.18

Batch Zn^{2+} adsorption by kaolin kinetic experimental data (Francisco & Sen, 2009)

Time, t (hr)	Amount of adsorption, q_t (mg/g)	Initial metal ion, Zn^{2+} concentration, mg/L
0	0.0	5.0
1	0.8771	
2	1.0479	
3	1.1036	
4	1.1138	
5	1.1139	
0	0	10.0
1	1.6465	
2	2.0396	
3	2.0625	
4	2.0779	
5	2.0778	

a. Explain the effect of initial metal ion concentration on adsorption kinetics, and
b. Justify the suitability of different kinetic models and find out adsorption capacity of adsorbent.

Solution:

(a) Effect of initial metal ion concentration. Plot q_t vs. time, t at two different initial metal ion concentrations which is shown in Figure 3.23.
From Figure 3.23, it was found that amount of adsorption increases with initial metal ion concentration. Initial speed of response is fast followed by slow intramolecular diffusion. It is a two-stage process.
(b) Plot log (q_e-q_t) vs. time t and check linearity. The experimental data do not fit well and produce very poor linear regression coefficient, R^2 value of 0.82 for which fitted linear plot is not presented here. This low value of R^2 indicates inapplicability of PFO model here. Then tested applicability of PSO model—Plot t/q_t vs. time, t, and check the linearity which is shown in Figure 3.24.

So, from Figure 3.24, it follows PSO kinetic model with higher linear regression coefficient (R) of above 0.97 at various initial metal ion concentrations. The theoretical, maximum adsorption capacity, q_e (mg/g), at a particular initial metal ion concentration can be determined from the slope of the straight line, and the second-order rate constant k_2 can be determined from the intercept of the straight line. Do this exercise on your own.

Table 3.19 presented the application of best fitted PSO model or PFO model equations and their model fitted kinetic parameters on selected metal ion and dye/antibiotic adsorption at solid/liquid interface by various adsorbents under specific process conditions reported by various researchers.

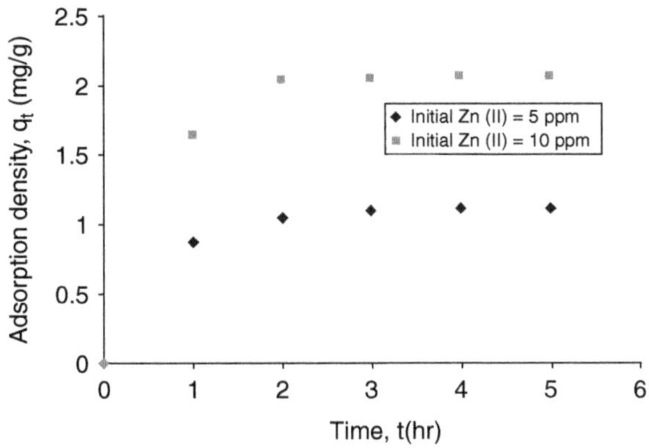

FIGURE 3.23
Variation of amount of adsorption, q_t (mg/g), with time, t, at two different initial metal ion concentrations

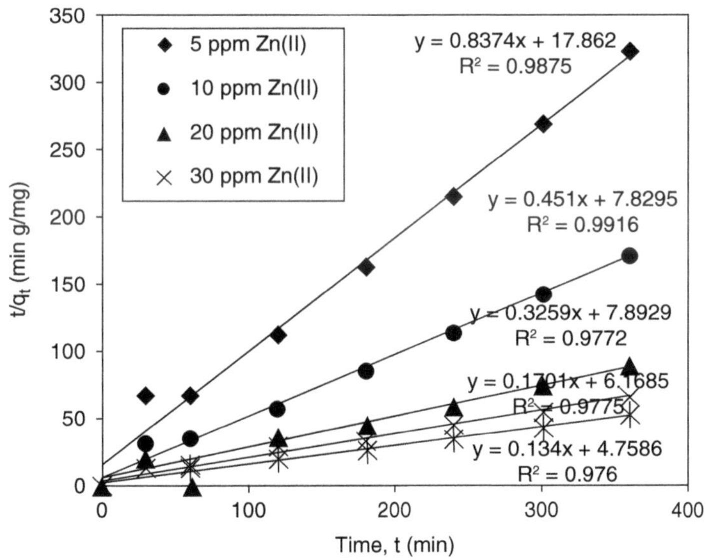

FIGURE 3.24
Linear fitting of PSO model to the batch kinetic experimental data

3.4.1.3 Weber-Morris Intraparticle Diffusion (IPD) Model

The IPD model, developed by Weber and Morris (Weber & Morris, 1963), is presented in Eq. 3.52. It is used to identify the rate-controlling step in the adsorption process and the mechanism of adsorption. The mechanism involves multiple steps: the solute adsorbate migrates from the bulk solution phase to the surface of the solid adsorbent, then diffuses through the

TABLE 3.19

Presented the application of best fitted PSO model or PFO model equations and their model fitted kinetic parameters on selected metal ion and dye adsorption at solid/liquid interface by various adsorbents under specific process conditions reported by various researchers

Adsorbate	Adsorbent	Application of best fitted kinetic model (PFO or PSO)	k_1 for PFO (min^{-1})	k_2 for PSO (g/ mg-min)	h for PSO (mg/g. min)	R^2	Reference
MB dye	JA-MIONs (magnetic	PFO and PSO	0.012	0.003	-	0.949 (PFO) and	Ghosh et al. (2023)
TCH	nano-iron oxides	both	0.007	0.009	-	0.937 for PSO	Ghosh et al. (2023)
antibiotic	modified with leaf	PFO and PSO	-	0.0104	1.29	0.99 for PFO and	Biswas et al. (2019b)
Zn^{2+}	extract)	both	-	2.42×10^{-3}	4.83	0.98 for PSO	Biswas (2020d)
MB Dye	Same JA-MIONs	Only PSO	-	4.03×10^{-4}	0.435	0.99	Biswas
Cu^{2+}	CABC (Calcium-aliginate	Only PSO	-	1.66×10^{-4}	1.44	0.99	et al. (2020b). Biswas,
Ni^{2+}	biochar)	Only PSO	-	6.65×10^{-3}	0.752	0.99	(2020d)
Cd^{2+}	CASB (Calcium aliginate	Only PSO	-	0.0018	0.182	0.99	Biswas et al. (2019c),
Phenol	sugarcane biochar	Only PSO	0.0457	0.0140	-	0.99	Biswas (2020d)
Hg^{2+}	Aliginate-biochar hybrid	Only PSO	0.246	0.038	-	0.98	Biswas et al. (2019c),
Pb^{2+}	Same aliginate-biochar	PFO and PSO	-	0.0024	-	0.986 for PFO and	Biswas (2020d)
Herbicide	composite	both	7.26×10^{-3}	2.89×10^{-2}	-	0.994 for PSO	Biswas (2020d)
MB dye	CABC & CASB	PFO and PSO	7.10×10^{-3}	2.30×10^{-2}	-	0.939 for PFO and	Al-Yaari et al. (2021)
MB dye	CASB	both	0.224	0.009	-	0.999 for PSO	Al-Yaari et al. (2023)
Cr^{6+}	Polymer-carbon-nanofibre	PSO				0.961	Samanth et. al (2024)
	PEI/GO (polymeric	PFO and PSO				0.926 for PFO and	Ghazy et al. (2024)
	graphine oxide	both				0.998 for PSO	Ghazy et al. (2024)
	composite)	PFO and PSO				0.954 for PFO and	Tian et al. (2024)
	Magnetic activated carbon	both				0.958 for PSO	
	PAN-nanofibres	PFO and PSO				0.903 for PFO and	
	PAN/SiO_2	both				0.917 for PSO	
	Sodium alginate biochar						

boundary layer to the surface of the adsorbent, followed by IPD into the interior of the sorbent (Konicki et al., 2013).

$$q_t = k_{id}t^{0.5} + C \tag{3.52}$$

where k_{id} is the IPD rate constant $\left[mg/(g \cdot min^{0.5})\right]$ and C represents the initial adsorption (mg / g), which also indicates the boundary layer thickness during diffusional mass transfer. A plot of q_t vs. $t^{0.5}$ should be linear, with k_{id} and C corresponding to the slope and intercept of the linear plot, respectively.

According to this model, a system is completely governed by IPD if the plot between q_t and $t^{0.5}$ produces a straight line passing through the origin. In most solid/liquid adsorption systems, adsorption kinetics is a multistep process; therefore, q_t vs. $t^{0.5}$ does not produce a single straight line passing through the origin. Instead, it often produces multiple straight lines, indicating that the system is not solely governed by IPD but also involves a combination of multiple diffusional steps (Biswas et al., 2019a, 2019b, 2019c; 2020a, 2020b; Dawood & Sen, 2012).

The overall adsorption process may be controlled either by one or more mass transfer steps such as film diffusion, external diffusion, surface diffusion, pore diffusion, or a combination of these (Biswas et al., 2020b, d). For example, Biswas et al. (2019b) reported batch Zn^{2+} adsorption kinetics results using a calcium alginate biochar adsorbent (CABC), and the results were fitted with the Weber-Morris IPD model (Eq. 3.52), for which a plot of q_t vs. $t^{0.5}$ is presented in Figure 3.25.

Figure 3.25 does not represent a single linear plot for the full time. However, it is clear from Figure 3.25 that there are three distinct linear sections for the full kinetic plot. Thus, the adsorption process consists of multiple steps: an initial rapid period, followed by a slower phase, and a stagnant zone. This trend indicates a multi-stage adsorption process where the solute adsorbate migrates from the bulk phase to the adsorbent surface through external mass transfer, followed by IPD (Dawood et al., 2014). The overall rate of adsorption is controlled by the slowest step, which could be either film diffusion or pore diffusion.

3.5 Continuous Mode of Adsorption Operation and Design Parameter Calculations

The continuous mode of adsorption using a fixed bed column has been widely applied for wastewater treatment. Generally, batch mode adsorption experiments are useful for identifying various process parameters and

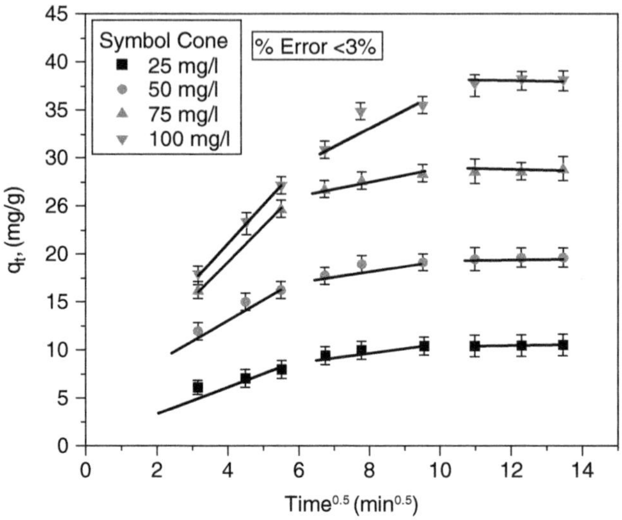

FIGURE 3.25
Weber-Morris (intraparticle diffusion) model application for adsorption of Zn^{2+} on CABC at different initial Zn^{2+} concentrations

Source: Taken from Biswas et al., 2019b, with written permission.

understanding their effects on the adsorption system. Batch kinetic and isotherm adsorption experiments are also essential to determine the optimum values of process parameters and the equilibrium time. Moreover, these experiments provide preliminary information about the effectiveness of selected adsorbents and their maximum adsorption capacity for a particular system under specified operating conditions.

However, the data obtained under batch conditions are generally not applicable to continuous adsorptive systems, where contact time is often insufficient to reach equilibrium (Afroze & Sen, 2018). Therefore, continuous adsorption studies are necessary to perform equilibrium studies using columns and to scale up for industrial applications. Various continuous modes of adsorption systems are currently operating at large scale, including packed bed operations, fluidized bed or semi-fluidized bed operations, and bubble column operations. Among these, fixed bed or packed bed adsorption is very common in real-world applications. Batch and fixed bed adsorption modes are typically used due to their ease of operation and maintenance (Patel, 2019; Patel, 2021). Fixed bed operations are more practical for treating large volumes of wastewater in industrial scenarios (Sazali et al., 2020).

3.5.1 Laboratory Fixed Bed or Packed Bed Adsorption Experimental Column Design and Procedure, and Column Dynamic Behaviour

Most industrial gas adsorption operations, as well as liquid adsorption operations, are carried out in a fixed bed of solid adsorbent, following the adsorption regeneration strategy (Dutta, 2007). The rate of adsorption in a bed, or part thereof, depends on several factors. Broadly, there are two mass transfer resistances in series: (i) the resistance to transport of the solute from the bulk of the gas or liquid to the surface of an adsorbent particle, presumably through a stagnant film of the gas or liquid, and (ii) the resistance due to the diffusion of the solute through the pores (macro-, meso-, and micro-pores) of an adsorbent particle to reach the adsorption sites within the particle (Dutta, 2007). Afroze et al. (2016b) conducted an up-flow fixed bed MB dye adsorption experiment using a Perspex glass column with a height of 30 cm and an internal diameter of 2.5 cm, as shown in Figure 3.26. A known amount of eucalyptus bark material ($\leq 106\ \mu m$) was packed inside the column to yield the desired bed height of the adsorbent. The top and bottom of the bed were covered by a layer of pre-equilibrated glass wool to prevent the loss of

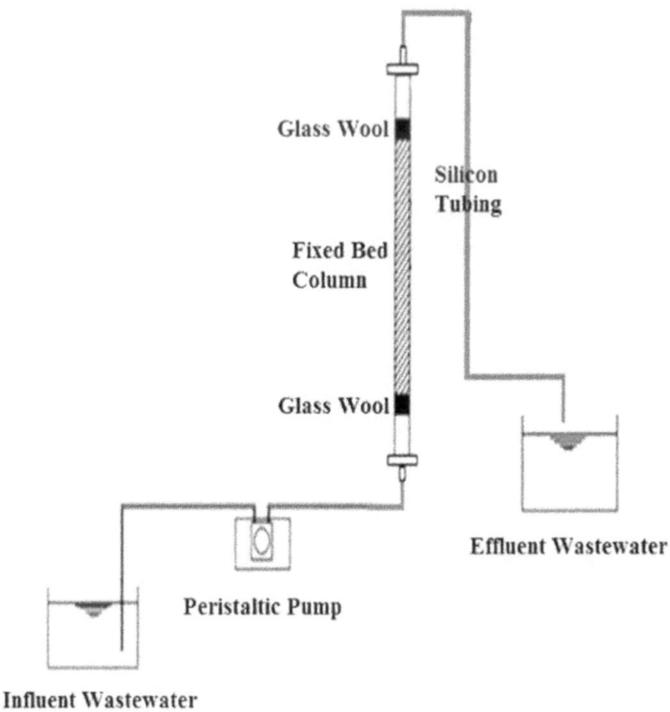

FIGURE 3.26
Schematic diagram of an up-flow fixed or packed bed adsorption column

Source: Afroze et al., 2016b with written permission.

adsorbent and to ensure a closely packed arrangement. The adsorbents were supported by two supporting layers of pre-equilibrated glass wool to ensure a securely packed arrangement (Figure 3.26). A porous sheet was attached at the bottom of the column to support the adsorbent bed and ensure uniform inlet flow and homogeneous liquid distribution into the column. A thin layer (~ 2 mm) of sand was placed at the top of the adsorbents to avoid bed fluidization due to the fine particle size of the adsorbent. Circular caps were used to enclose both ends of the column (Afroze et al., 2016). A variable speed peristaltic pump was used to introduce the influent dye solution at the desired flow rate through the column (Figure 3.26). The contaminated wastewater was pumped upwards to avoid channelling due to gravity. The experiments were carried out to study the effects of influent flow rates, bed height, and initial adsorbate concentration on the column dynamics through breakthrough curves (BTCs). The effluent was collected at regular time intervals, and the concentration of adsorbate was measured. The obtained BTCs under various operating conditions were then fitted with various column kinetic models, and model parameters were determined.

Figure 3.27 shows a down-flow fixed bed adsorption column operation. Downward flow can be realized by hydrostatic pressure alone, which helps keep the media in the bed. Generally, down-flow operation can achieve higher flow rates without fluidizing the bed (Kamal, 2020).

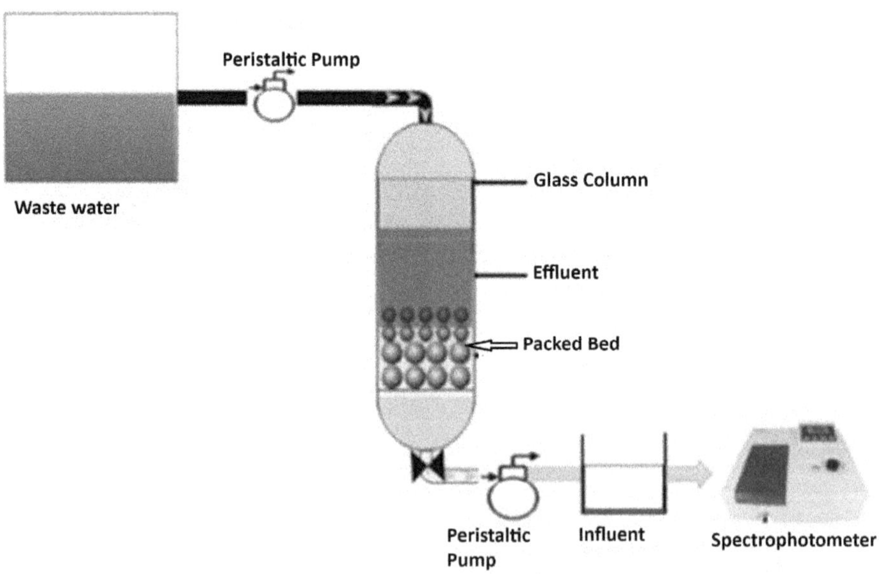

FIGURE 3.27
Schematic diagram of a down-flow fixed or packed bed adsorption column (Courtesy: Kamal, 2020)

3.5.2 Dynamic Modelling and Column Adsorption Operation Analysis

3.5.2.1 Theory of BTC

The performance of a fixed bed column was analysed using the theory of the BTC. The time for breakthrough appearance and the shape of the BTC are very important characteristics for determining the operation and the dynamic response of an adsorption column (Yagub et al., 2014b). The shape of the concentration-time profile of the BTC is also critical for understanding the adsorption dynamic response and designing the adsorption column (Afroze et al., 2016b; Yaqub et al., 2014b). The pattern of BTCs largely depends on factors such as flow rate, dye concentration, bed height, and column dimension.

In adsorption studies, the BTC often observed is "S" shaped. The shorter the dynamic breakthrough zone in the column, the longer the column service time and the larger the fully utilized sorbent portion inside the column (Afroze & Sen, 2018). A typical plot of C_t / C_0 vs. treated volume (V) or service time (t) for a given bed height, where C_t is the effluent concentration (mg/ L) at time t, C_0 is the influent (initial) concentration (mg/ L), and t is the flow time (min), is called the breakthrough curve (BTC), as presented in Figure 3.28 (Afroze et al., 2016b; Dutta, 2007).

The concentration at the breakthrough point (C_b) is chosen arbitrarily at a low value. The effluent concentration (C_t) from the column that reaches about 0.1% of the influent concentration (C_0) is considered the breakthrough point. When the effluent concentration (C_t) approaches 99.5% of the influent adsorbate concentration (C_0), the adsorbent is considered essentially exhausted (Faust and Aly, 2013). The area under the BTC represents the total quantity of sorbed adsorbate for a given feed concentration. The effectiveness of an adsorbent can be determined from the BTC of the effluent concentration, where a typical S-shaped BTC is commonly observed (Cruz-Olivares, Pérez-Alonso et al., 2013).

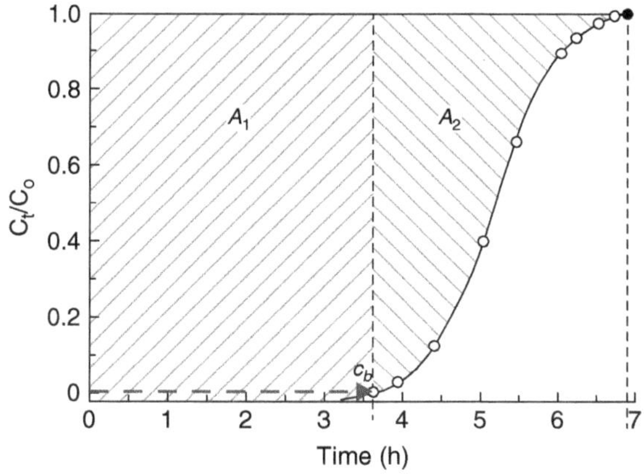

FIGURE 3.28
Ideal breakthrough curve

Source: Afroze et al., 2016b with written permission; Dutta, 2007; Geankoplis, 2003.

The time equivalent to total or stoichiometric capacity is calculated using the following equation (Yagub et al., 2014a):

$$t_r = \int_{t=0}^{t=\infty} \left(1 - \frac{c_t}{c_0}\right) dt = A_1 + A_2 \tag{3.53}$$

The time equivalent to usable capacity is determined using Eq. 3.54 (Yagub et al., 2014a):

$$t_u = \int_{t=0}^{t=t_b} \left(1 - \frac{c_t}{c_0}\right) dt = A_1 \tag{3.54}$$

The usable capacity of the bed up to the breakthrough time point, t_b, and area, A_2, allows calculation of the height of the unused bed (HUNB).

$$t_u \approx t_b \tag{3.55}$$

The area under the curve as per Eq. 3.53 gives the total time value, while the area under the curve as per Eq. 3.54 gives the usable capacity time. The area under the curve can be determined using graphical methods or by numerical integration. The fraction of the total bed capacity or length utilized up to the break point is expressed as $\frac{t_b}{t_r}$. The HUNB or mass transfer zone (MTZ) is calculated using Eqs. 3.56 and 3.57 (Yagub et al., 2014a):

$$H_{\text{UNB}} = \left(1 - \frac{t_u}{t_t}\right) H_T = \left(1 - \frac{t_b}{t_t}\right) H_T \tag{3.56}$$

$$MTZ = H_{\text{UNB}} \tag{3.57}$$

where H_T is the total bed height (cm).

The used bed length up to the break point H_B is calculated using Eq. 3.58 (Yagub et al., 2014a):

$$H_B = \left(\frac{t_b}{t_t}\right) H_T \tag{3.58}$$

The total effluent volume V_{eff} is estimated using Eq. 3.59 (Yagub et al., 2014a):

$$V_{\text{eff}} = Q \cdot t_{\text{total}} \tag{3.59}$$

where Q is the volumetric flow rate (mL/min) and t_{total} is the total flow time (min).

The BTC plot can be used to find the total quantity of adsorbate adsorbed. The total amount of adsorbate adsorbed in milligrams for a given inlet concentration is determined using Eq. 3.60 (Yagub et al., 2014a):

$$q_{\text{total}} = \frac{QA}{1000} = \left(\frac{QC_0}{1000}\right)\int_{t=0}^{t=t_{\text{total}}} \frac{C_t}{C_0}\,dt \tag{3.60}$$

The total amount of adsorbate contaminant pumped into the adsorption separation column (m_{total}) is calculated using Eq. 3.60 (Ataei-Germi and Nematollahzadeh, 2016):

$$m_{\text{total}} = \frac{C_0 \cdot Q \cdot t_{\text{total}}}{1000} \tag{3.61}$$

The total percentage removal of adsorbate MB dye can be calculated from the ratio of the total quantity of MB adsorbed (q_{total}) to the total amount pumped into the column (m_{total}), as shown in Eq. 3.61 (AtaeiGermi and Nematollahzadeh, 2016):

$$\%\ \text{Removal} = \left(\frac{q_{\text{total}}}{m_{\text{total}}}\right)\times 100 \tag{3.62}$$

Sample Calculation for t_r (Numerical Integration)

$$t_r = \int_0^{\infty}\left(1-\frac{C}{C_0}\right)dt$$

t(min)	C/C_0	$1-C/C_0$	Δt	$(1-C/C_0)_{avg}$	$\Delta t \times (1-C/C_0)_{avg}$
141	0	1	13	1	13
154	0	1	12.7	0.991	12.58
166.7	0.018	0.982	23	0.919	21.137
189.7	0.144	0.856	15.3	0.8165	12.49
205	0.223	0.777	20.6	0.683	14.069
225.6	0.411	0.589	20.4	0.501	10.220
246	0.587	0.413	15	0.3605	5.4
261	0.692	0.308	21	0.2505	5.260
282	0.807	0.193	15	0.1495	2.2425
297	0.894	0.106	21	0.07	1.47
318	0.966	0.034	20	0.022	0.44
338	0.99	0.01	12	0.005	0.06
350	1	0			

Total= 98.366 min

This example shows how to calculate the area $A = \int_0^\infty \left(1 - \frac{C}{C_0}\right) dt$ by numerical integration using tabular data of C / C_0 vs. time (t).

3.5.2.2 Fixed Bed or Packed Bed Column Adsorption Kinetic Models

Kinetic modelling of fixed bed column is required to predict the dynamic behaviour of the bed in an industrial scale operation. Thomas model, Yoon-Nelson model, Clark constants, and Bed Depth Service Time (BDST) model are the most used kinetic models to analyse the column performance. These models are used to calculate the column kinetics parameters and the maximum adsorption capacity of the fixed bed column.

3.5.2.2.1 Thomas Dynamic Model

The Thomas model (Thomas, 1944) is one of the most widely used models in describing the column performance and prediction of BTCs. It assumes a plug flow behaviour in the bed, and the rate driving force in adsorption follows the second-order reversible kinetics reaction and Langmuir isotherm model (Dawood et al., 2018; Xu et al., 2013). The model can be represented by Eq. 3.62:

$$\frac{C_t}{C_0} = \frac{1}{1 + \exp\left[\left(\frac{K_T}{Q}\right)(q_0 m - C_0 Q t)\right]} \tag{3.62}$$

The kinetic coefficient, K_T, Thomas rate constant, and the adsorption capacity of adsorbent, q_0, can be determined from a plot of C_t/C_0 against t for a given flow rate using non-linear regression analysis.

However, the linearized form of the model can also be represented by Eq. 3.63 (Dawood et al., 2018; Yagub et al., 2014b).

$$\ln\left(\frac{C_0}{C_t} - 1\right) = \left(\frac{K_T q_0 m}{Q}\right) - \left(\frac{K_T C_0 V}{Q}\right) \tag{3.63}$$

where K_T is Thomas rate constant (mL/mg min), q_o is the adsorption capacity (mg/g), m is the amount of adsorbent in column (g), C_0 *is* the inlet adsorbate concentration (mg/L), C_t is the effluent adsorbate concentration (mg/L), V is total volume (mL), and Q is inlet flow rate (ml/min). The kinetic coefficient K_T and q_0 can also be calculated from the intercept and slope of the linear plot ln [(C_0/C_t)–1] against t at a given flow rate.

3.5.2.2.2 *Yoon-Nelson Model*

The Yoon and Nelson model (Yoon & Nelson, 1984) is judged as the less complicated column model because it does not require detailed data concerning the characteristics of adsorbate, the type of adsorbent, and the physical properties of adsorption bed (Afroze & Sen, 2018).

The linearized form of this model can be represented by Eq. 3.64 (Afroze & Sen, 2018; Dawood et al., 2018; Yagub, Sen et al., 2015).

$$\ln\left(\frac{C}{C_o - C}\right) = K_{YN}t - K_{YN}\tau \tag{3.64}$$

where K_{YN} is Yoon and Nelson rate constant (min^{-1}), Co is the inlet adsorbate concentration (mg/L), C is the effluent adsorbate concentration (mg/L), t is the breakthrough time (min), and τ is the time required for 50% adsorbate breakthrough (min). The values for K_{YN} and $K_{YN}\tau$ are obtained from the slope and intercept of the linear plot $\ln\left(\frac{C}{C_0 - C}\right)$ against time (t).

3.5.2.2.3 *Bed Depth Service Time (BDST) Model*

BDST is a simple model which states that column bed depth and service time bears a linear relationship in terms of process concentration and adsorption parameters. The BDST model was derived from the equation described by Adams–Bohart but was modified by Hutchins (Hutchins, 1973). It models a linear relationship between the column bed depth and the service time. Service time is the time required for an adsorbent to adsorb a specific amount of adsorbate from a solution before regeneration is needed (Afroze et al., 2016b; Dawood et al., 2018).

The BDST model equation can be represented by the following Eq. 3.65 (Dawood et al., 2018; Afroze & Sen, 2018).

$$t = \left(\frac{N_0 H_T}{C_0 u}\right) - \left(\frac{1}{K_0 C_0}\right)\ln\left(\frac{C_0}{C_t} - 1\right) \tag{3.65}$$

where *Co* is the inlet solute concentration (mg/L), C_t is effluent concentration of solute in the liquid phase (mg/L), u is influent linear velocity (cm/min), N_0 is the adsorption capacity (mg/L), K_0 is the rate constant (L/mg min), t is time (min), and H_T is the bed height of column (cm).

A plot of t vs. bed height H_T should yield a straight line where N_0 and K_0 can be determined from slope and intercept of the plot. The BDST model can be further simplified into Eq. 3.66 as follows (Afroze & Sen, 2018; Dawood et al., 2018).

$$t = az + b \tag{3.66}$$

where $a = \frac{N_0}{C_0 U}$ is the slope and $b = -\left(\frac{1}{K_0 C_0}\right)\ln\left(\frac{C_0}{C_t} - 1\right)$ is the intercept.

Adsorption capacity of each model for different operational process parameters such as initial concentration of adsorbate, bed height, and flow rate is calculated and mentioned by various researchers for designing the adsorption column (Patel, 2019). Table 3.20 depicts the equation, plot, and parameters of each model. It also mentioned the variation, i.e., increase or decrease in column adsorption model parameters with respect to increase in operation parameters.

Error Calculation

To validate the various dynamic models, scientists use validation parameters based on the linear regression coefficient (R^2) value, which is confirmed by least square error (SS) calculation. The following SS calculation equation is used for various column model validations (Dawood et al., 2018):

$$SS = \frac{\sum\left[\left(\frac{C_t}{C_0}\right)_{\text{cal}} - \left(\frac{C_t}{C_0}\right)_{\text{Exp}}\right]^2}{N} \tag{3.67}$$

where $\left(\frac{C_t}{C_0}\right)_{\text{cal}}$ is the theoretical normalized concentration, $\left(\frac{C_t}{C_0}\right)_{\text{Exp}}$ is the experimental value, and N is the total number of experimental data points.

Table 3.21 presents compiled results of inorganics and organics adsorption from the aqueous phase through fixed bed adsorption column operations using various fixed bed adsorbent materials over the last few years.

3.6 Application of Fixed/Packed Bed vs. Fluidized Bed in Continuous Wastewater Treatment Operation

Fixed and fluidized beds are two conventional systems for gas-liquid-solid or gas-solid or liquid-solid adsorptive separation processes. Both have advantages and disadvantages. The packed bed has a low axial dispersion coefficient, resulting in a high conversion rate, and there is no back mixing, but it has a high pressure drop. On the other hand, in the fluidized bed, there is no channelling or flooding, and the pressure drop is lower than packed bed operation, but the high dispersion and back mixing reduce the conversion rate. However, both systems have been widely used in various chemical and biochemical engineering operations (Biswas, 2020d).

TABLE 3.20

Details of most commonly used adsorption models and its parameter variation in column adsorption model w.r.t. increasing operation parameters

Column adsorption model	Linear equation	Plot	Parameter of model	Operation parameters			
				Initial conc.	Flow rate	Bed height	Temperature
Thomas model (TM)	$\ln\left[\left(\frac{C_O}{C_t}\right)-1\right]=\frac{k_{TH}q_0m}{Q}-\frac{k_{TH}C_oV_{eff}}{Q}$	$\ln\left[(C_o/C_t)-1\right]$ vs. time	k_{TH} q_0	Decreased Increased	Increased Decreased	Decreased Increased	Decreased Increased
Bed depth service time model (BDST)	$t=\frac{N_O}{C_0F}Z-\frac{1}{k_{BDST}C_0}\ln\left(\frac{C_0}{C_t}-1\right)$	Bed height vs. time	N_o k_{BDST}	Increased Increased	Increased Increased	N.A. N.A.	Increased Increased
Adam and Bohart model (ABM)	$\ln\left[\left(\frac{c_0}{c_t}\right)-1\right]=K_{AB}N_O\frac{Z}{u}-K_{AB}C_tt$	$\ln\left[(C_o/C_t)-1\right]$ vs. time	k_{AB} N_o	Increased Decreased	Increased Decreased	Decreased Increased	Decreased Increased
Yoon-Nelson model (YNM)	$\ln\left(\frac{C_t}{C_o-C_t}\right)=k_{YN}t-\tau k_{YN}$ $q_{OYN}=\frac{q_{(total)}}{X}=\frac{C_0Q_\tau}{1000X}=\frac{\frac{1}{2}C_O\left[(Q/1000)X2\tau\right]}{X}$	$\ln\left(\frac{C_t}{C_0-C_t}\right)_{versus}$ time	k_{YN} T	Increased Decreased	Increased Decreased	Decreased Increased	N.A. N.A.
Clark model (CM)	$\frac{C_t}{C_o}=\frac{1}{\left(1+Ae^{-n}\right)^{1}(n-1)}$	C_t/C_0 vs. time	A	Decreased	Decreased	Increased	Increased

(Coutinued)

TABLE 3.20
(Coutinued)

Column adsorption model	Linear equation	Plot	Parameter of model	Operation parameters			
				Initial conc.	Flow rate	Bed height	Temperature
			R	Increased	Increased	Decreased	Decreased
Wolborska model (WM)	$\ln\left(\frac{C_i}{C_o}\right)=\frac{\beta C_o t}{N_o}-\frac{\beta Z}{U}$	$\ln(C_t/C_0)$ vs. time	B	Decreased	Increased	Decreased	Decreased
			N_o	Increased	Decreased	Increased	Increased
Modified dose response model (MDRM)	$\ln\left(\frac{C_t}{C_0-C_t}\right)=a\ln(C_0Qt)-a\ln(q_{mdr}m)$	$\ln\left(\frac{C_t}{C_0-C_t}\right)$ vs. $\ln(C_0Q_t)$	A	Decreased	Increased	Increased	N.A.
			B	Increased	Increased	Increased	N.A.

TABLE 3.21

Few compilation results on the adsorption of inorganics and organics by various adsorbents in a continuous fixed bed column operation (modified and updated from Afroze & Sen, 2018)

Adsorbents	Adsorbates (heavy metals and dyes)	Column process/conditions	Maximum adsorption capacity, q_m (mg g^{-1})	References
Chitosan magnetic hydrogel beads (ChM)	Arsenic oil and grease	Bed height: 13 cm, flow rate: 13 mL hr^{-1} and arsenate concentration 5 mg L^{-1}	104 (Thomas)	Mendizabal et al. (2023)
Rice husks waste-based adsorbents	perchlorate	Bed height: 100 cm, flow rate: 10.5 mL min^{-1} and adsorbate concentration 200 mg L^{-1}	5513.40 mg L^{-1} (BDS)	Umembamalu et al. (2020)
Coconut shell-based activated carbon	Pb^{2+}, Cd^{2+}	Bed height: 50 cm, flow rate: 50 mL min^{-1} and Perchlorate concentration 100 mg L^{-1}	21.5 (Thomas)	Radhika et al. (2018)
Magnetic polymer nanocomposite	MB dye	Bed height: 20 cm, fixed flow rate and Fixed initial metal ion concentration	36.20 for Pb^{2+} (Thomas) 37.25 for Cd^{2+} (Thomas)	Zendehdel et al. (2018)
Pinecone +kaolin	Phenol	Bed height: 5 cm, flow rate: 13 mL min^{-1} and MB concentration = 100 mg L^{-1}	16.9 (Thomas)	Dawood et al. (2018)
Crab eye activated carbon (CEAC)	Pb^{2+}, Zn^{2+}	Bed height: 15 cm, fixed flow rate: 5 mL min^{-1} and Initial phenol concentration = 100 mg L^{-1}	342.82 (Thomas)	Lallan et al. (2017)
Modified beer lees (MBL)	Ni^{2+} and Co^{2+}	Bed height: 3 cm, flow rate: 50 mL min^{-1} and Metal concentration 30 mg L^{-1} (Pb^{2+}), 15 mg L^{-1}(Zn^{2+})	29.6 (Pb^{2+}), 5.43(Zn^{2+})	Dong and Lin (2017)
Tectona grandis leaves (raw)		Bed height: 1.0 – 3.0 cm, flow rate: 1 – 5 mL min^{-1} and metal ion concentration: 25 – 75 mg L^{-1}	19.48 (Ni^{2+}) and 20.66 (Co^{2+})	Vilvanathan and Shanthakumar (2017)
Vanda species (*Bassia latifolia* tree biomass)	Pb^{2+}	Amount of biomass: 60 g, inlet Pb^{2+} solution flow rate: 13 mL min^{-1} with initial concentration: 100 mg L^{-1}	83.33	Sahu et al. (2017)

(Coutinued)

TABLE 3.13
(Coutinued)

Adsorbents	Adsorbates (heavy metals and dyes)	Column process/conditions	Maximum adsorption capacity, q_m (mg g^{-1})	References
Tea waste, maple leaves, and mandarin peel (combined)	Cd^{2+}, Cu^{2+}, Pb^{2+} and Zn^{2+}	Optimum conditions: 31 cm adsorbent bed height, 10 mL min^{-1} flow rate and 20 mg L^{-1} initial concentration	38.25, 63.37, 108.12 and 35.23 for Cd^{2+}, Cu^{2+}, Pb^{2+} and Zn^{2+} respectively	Abdolali et al. (2017)
Eucalyptus sheathiana bark biomass	Methylene blue (MB)	MB flow rate: 10–15 mL min^{-1}, initial MB concentration: 50–100 mg L^{-1} and adsorbent bed height: 10–15 cm	48% removal	Afroze et al. 2016b)
Saw dust (magnetized)	Cu^{2+} and Ni^{2+}	Bed height: 2.0 – 8.0 cm, flow rate: 15 – 25 mL min^{-1} and metal ion concentration: 10 – 30 mg L^{-1}	29.42 (Cu^{2+}) and 20.31(Ni^{2+})	Kapur and Mondal (2016)
Rice husk	Cd^{2+}	initial concentration: 10 – 50 mg L^{-1}, flow rate: 10–90 mL min^{-1}, adsorbent bed height: 30–40 cm	69% removal	Kulkarni and Kaware (2015)
Corn cob	Zn^{2+}, Cr^{6+}, Ni^{2+}	The range pH of 5–8 and contact time from 0.25 – 5 hrs	91.35% (Zn^{2+}), 95.05% (Cr^{6+}), 99.09% (Ni^{2+}) removal efficiency	Ali et al. (2014)
Agaricus bisporus mushroom (sodium hydroxide modified)	Pb^{2+}	Optimum conditions for maximum adsorption: 4 cm of adsorbent bed height, flow rate of 1.0 mL min^{-1} with metal ion concentration of 50 mg L^{-1}	60.8	Long et al. (2014)
Rice straw	Cd^{2+} and Pb^{2+}	Optimum conditions: adsorbent bed height: 20 cm, flow rate: 10 mL min^{-1} and initial metal ion concentration: 0.01 mmol L^{-1}	0.076 (Cd^{2+}) and 0.151 (Pb^{2+})	Soetaredjo et al. (2014)
Palm fibre-derived activated carbon	Pb^{2+}	Initial concentration: 50 – 150 mg L^{-1}, flow rate: 5–10 mL min^{-1} and adsorbent bed height: 50–150 mm	8.754 (at flow rate of 5 mL min^{-1} and inlet metal concentration of 100 mg L^{-1}).	Nwabanne and Igbokwe (2012)

Source: Patel, 2019.

Fluidized bed adsorption column operation in wastewater treatment has been reported by many researchers (Biswas, 2020d). Adsorption modelling in a liquid/solid fluidized bed has been investigated and it has been reported that this kind of bed has a shorter breakthrough time for pollutant removal (Veeraraghavan, Fan, and Mathews, 1989). In another study, activated carbon was prepared from apricot stone and used in a fluidized bed adsorption column operation for the removal of divalent heavy metals like Pb, Cu, Cd, and Zn (Biswas, 2020d). Correa et al. (2007) developed a fluidized bed system for the adsorptive removal of aqueous-phase phenol with commercial macro-porous resin (Corrêa, Calçada, and Peçanha, 2007). Fluidized bed adsorption columns have also been successfully used for the removal of various air pollutants. A model was proposed for the removal of CO_2 in a lab-scale fluidized bed system (Yaghoobi-Khankhajeh, Alizadeh, and Zarghami, 2018). The application of fluidized bed bioreactors in various environmental engineering solutions, including wastewater treatment, has been reviewed by Özkaya et al. (2019).

3.7 Application of a Gas-Liquid-Solid Three-Phase Semi-fluidized Bed in Continuous Wastewater Treatment Operation

Semi-fluidization is a technique for liquid-solid or gas-liquid-solid contacting, similar to normal fluidized or packed bed systems, which was first introduced by Fan and co-workers in the early 60s to investigate the mass transfer between solid and liquid in a semi-fluidized bed system (Fan, Yang, and Wen, 1959). In this system, a packed bed at the top is followed by fluidized bed generation in the same reacting column (Biswas et al., 2019d). This bed can be formed by providing sufficient space for the bed material to freely expand and arrest the escaping particles by a top restrain. The degree of semi-fluidization is determined between the minimum semi-fluidization (when the first particle touches the top restrain) and maximum fluidization (when all the bed material reaches the top or forms an inverse packed bed) (Biswas et al., 2019; Fan & Wen, 1961).

Such a system can create substantial intimate contact between the gas, liquid, and solid particles and provide significant advantages for physical, chemical, and biological processing operations. The difficulties of fluidized beds, such as solid back-mixing, particle attrition, and erosion of surfaces, as well as issues in packed bed systems like non-uniform bed temperatures, solids isolation, and channelling, are reduced in a semi-fluidized bed system (Biswas et al., 2019d; Schubert & Larachi, 2009). Two models of gas-liquid-solid semi-fluidized bed operation have been reported in the literature: co-current and counter-current mode operations. The co-current three-phase

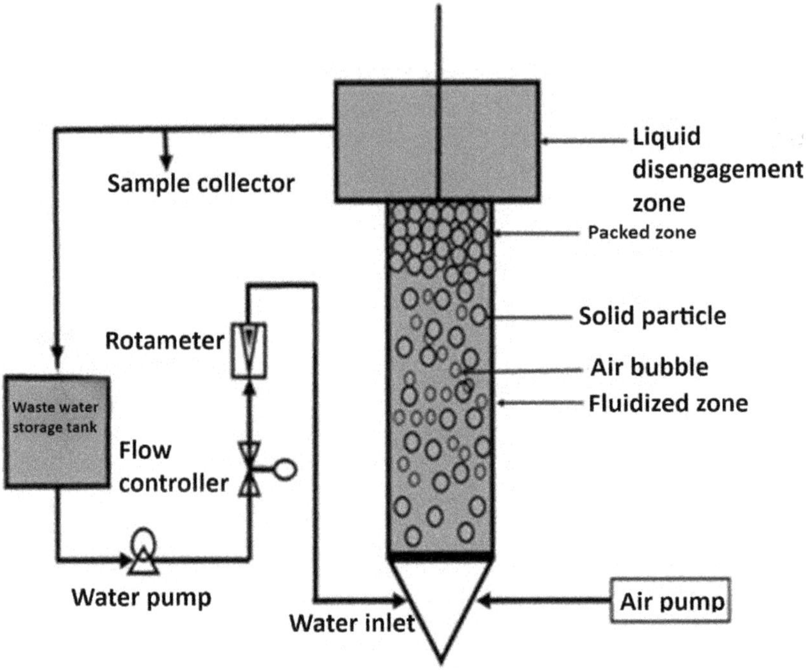

FIGURE 3.29
Schematic diagram of a gas-liquid-solid three-phase semi fluidized bed adsorption column for wastewater treatment (Biswas, 2020d; Biswas et al., 2019d with written permission)

system is the most prominent, with upward gas and liquid flow. However, this type of three-phase semi-fluidized bed may offer high pressure drops due to the presence of a packed section at the top, resulting in higher gas hold-up and a higher interfacial area in the bed (Biswas et al., 2019d; Jena et al., 2009).

A gas-liquid-solid three-phase semi-fluidized bed system involving relatively low density solids or a low solid/liquid density ratio was developed by Biswas et al. (2019d), for which schematic diagrams are shown in Figures 3.29 and 3.30. The adsorber column consists of three different sections: the gas-liquid distribution section, the main reaction or gas-liquid-solid contacting section, and the liquid disengagement section (Figures 3.29 and 3.30). The distributor section is 22 cm high with a conical shape divergence angle of 5.30°. A water inlet nozzle with a diameter of 0.126 cm and an outlet nozzle with a diameter of 10.03 cm are connected to the main reacting section by flanges. A brass plate with 90 holes of 2 mm diameter is attached between the flanges (Figure 3.30). An air connection enters this section to ensure proper mixing of air and water before entering the main column section. The main section is made of a Perspex tube with an inner diameter of 10 cm and a thickness of 3 mm. Pressure taps are connected at different

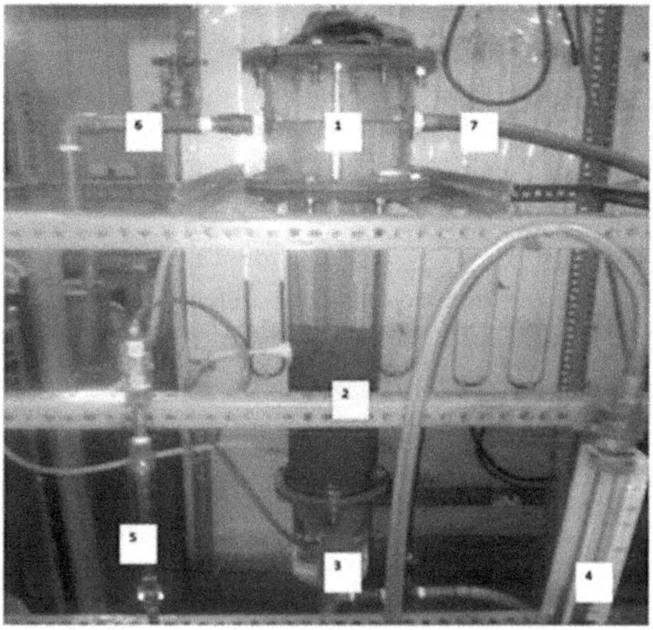

FIGURE 3.30
Pictorial view of a laboratory-scale gas-liquid-solid three-phase semi-fluidized bed adsorption column for wastewater treatment (Biswas, 2020d with written permission)

positions along the column to monitor pressure at different points. The top restraint is made of a stainless-steel plate with an outer diameter (OD) of 10 cm and 2 mm holes to restrict the upward movement of solid particles. The top section also contains a liquid disengagement section, where the liquid accumulates just after contact takes place. A sample collection point is also connected here (Biswas et al., 2019, 2020a, 2020b, 2020c, 2020d). Biswas et al. (2020a, 2020b, 2020c, 2020d) reported high removal efficiency for this gas-liquid-solid three-phase semi-fluidized bed adsorption column operation in the removal of inorganics/organics by CABC composite adsorbent from wastewater, including treatment of steel plant effluents.

Bibliography

Abdolali, A., H. H. Ngo, W. Guo, J. L. Zhou, J. Zhang, S. Liang, S. W. Chang, D. D. Nguyen, Y. Liu (2017). Application of a breakthrough biosorbent for removing heavy metalsfrom synthetic and real wastewaters in a lab– Scale continuous fixed– Bed column. *Bioresource Technology*, 229, 78–87.

Afroze S., T. K. Sen (2018). Removal of aqueous phase heavy metal ions and dyes using agricultural solid wastes adsorbents: A review. *Water, Air & Soil Pollution*, 229(225), 1–50. https://doi.org/10.1007/s11270-018-3869-z

Afroze, S. (2016a). Ph.D thesis entitled aqueous phase adsorption of organic/inorganic contaminants by eucalyptus bark (eucalyptus sheathiana) biomass. Curtin University, Perth, Australia. URI: https://hdl.handle.net/20.500.11937/1561

Afroza, S., T. K. Sen, M. Ang (2016). Adsorption removal of Zinc (II) from aqueous phase by Raw and Base Modified *Eucalyptus Sheathiana* Bark: Kinetics, mechanism and equilibrium study. *Process Safety and Environmental Protection*, 102, 336–357. https://doi.org/10.1016/j.psep.2016.04.009

Afroze, S., T. K. Sen, H. Ang (2016b). Adsorption performance of continuous fixed bed column for the removal of methylene blue (MB) dye using Eucalyptus *sheathiana* bark biomass. *Research on Chemical Intermediates*, 42(3), 2343–2364.

Akram, M., H. N. Bhatti, M. Iqbal, S. Noreen, S. Sadaf (2017). Biocomposite efficiency for Cr(VI) adsorption: Kinetic, equilibrium and thermodynamics studies. *Journal of Environmental Chemical Engineering*, 5(1), 400–411.

Ali, S., A. Khalid, R. Majid (2014). The removal of Zinc, Chromium and Nickel from industrial waste water using Corn cobs. *Journal Science*, 55, 121–131.

Argun, M. E., D. Güclü, M. Karatas (2014). Adsorption of Reactive Blue 114 dye by using a new adsorbent: Pomelo peel. *Journal of Industrial and Engineering Chemistry*, 20(3), 1079–1084.

Ataei-Germi, T., A. Nematollahzadeh (2016). Bimodal porous silica microspheres decorated with polydopamine nano-particles for the adsorption of methylene blue in fixed-bed columns. *Journal of Colloid and Interface Science*, 470, 172–182.

Azad, H., M. Mohsennia, C. Cheng, A. Amini (2021). Facile fabrication of PVB-PVA blend polymer nanocomposite for simultaneous removal of heavy metal ions from aqueous solutions: Kinetic, equilibrium, reusability and adsorption mechanism. *Journal of Environmental Chemical Engineering*, 9, 106214. https://doi.org/10.1016/j.jece.2021.106214

Barkakati, P., A. Begum, M. L. Das, P. G. Rao (2010). Adsorptive separation of Ginsenoside from aqueous solution by polymeric resins: Equilibrium, kinetic and thermodynamic studies. *Chemical Engineering Journal*, 161, 34–45. https://doi.org/10.1016/j.cej.2010.04.018

Bayuo, J., J. Mwemezi, Rwiza, M. S., K. M. Mte (2023). Removal of heavy metals from binary and multicomponent adsorption systems using various adsorbents – A systematic review. *RSC Advances*, 13, 13052–13093. https://doi.org/10.1039/d3ra01660a

Behrooz, B. A, S. Oustan, H. M. Hosseini, H. Etesami, E. Padoan G. Magnacca, et al. (2023). The importance of presoaking to improve the efficiency of MgCl2-modified and nonmodified biochar in the adsorption of cadmium. *Ecotoxicology and Environmental Safety*, 2257, 114932. https://doi.org/10.1016/j.ecoenv.2023.114932

Ben-Ali, S., I. Jaouali, S. Souissi-Najar, A. Ouederni (2017). Characterization and adsorption capacity of raw pomegranate peel biosorbent for copper removal. *Journal of Cleaner Production*, 142, 3809–3821.

Benjamin, M. M., J. Leckie (1981). Multiplesite adsorption of Cd, Cu, Zn and Pb on amorphous iron oxyhydroxide. *Journal of Colloid and Interface Science*, 79, 209–219.

Biswas, S., Manisha Bal, S. K. Behera, T. K. Sen, B. C. Meikap (2019a). Process optimization study of Zn2+ adsorption on biochar-alginate composite adsorbent by Response Surface Methodology (RSM). *Water,* 11(2), 325. https://doi.org/10.3390/w11020325

Biswas, S., T. K. Sen, A. M. Yeneneh, B. C. Meikap (2019b). Synthesis and characterization of a novel Ca-alginate-biochar composite as efficient zinc (Zn^{2+}) adsorbent: Thermodynamics, process design, mass transfer and isotherm modelling. *Separation Science and Technology,* 54(7), 1106–1124.

Biswas, S., B. C. Meikap, T. K. Sen (2019c). Adsorptive removal of aqueous phase copper (Cu2+) and nickel (Ni2+) metal ions by synthesize Biochar-biopolymeric hybrid adsorbents and process optimization by response surface methodology (RSM). *Water, Air & Soil Pollution,* 230, 197. https://doi.org/10.1007/s11270-019-4258-y

Biswas, S., T. K. Sen, B. C. Meikap (2019d). Experimental hydrodynamic and bed characteristics of co-current gas-liquid-solid three phase semifluidization with liquid as the continuous phase. *Particulate Science and Technology,* 38(7), 1–13.

Biswas, S., H. Siddiqi, B. C. Meikap, T. K. Sen, M. Khiadani (2020a). Preparation and characterization of raw and inorganic acid-activated pine cone biochar and its application in the removal of aqueous-phase Pb^{2+} metal ions by adsorption. *Water, Air, and Soil Pollution,* 231(1), 3.

Biswas, S., S. S. Mohapatra, U. Kumari, B. C. Meikap, T. K. Sen (2020b). Batch and continuous closed circuit semi-fluidized bed operation: Removal of MB dye using sugarcane bagasse biochar and alginate composite adsorbents. *Journal of Environmental Chemical Engineering,* 8(1), 103637. https://doi.org/10.1016/j.jece.2019.103637

Biswas, S., R. K. Diwakar, Ipshita D. Behera, B. C. Meikap, T. K. Sen, M. Khiadani (2020c). Aqueous phase phenol removal from synthetic and real steel plant effluents through a batch and Semifluidized bed clumn operation: Experimental and model analysis. *Journal of Environmental Chemical Engineering,* 8, 104441. https://doi.org/10.1016/j.jece.2020.104441

Biswas, S. (2020d). Ph.D thesis entitled Design and Development of a Three Phase Semifluidized Bed Reactor for Wastewater Treatment with Reference to Steel Plant Effluent. Curtin University, Perth, Australia and IIT-Kharagpur, India under collaborative Ph.D. program. https://hdl.handle.net/20.500.11937/82127

Boruah, P., A. Sarma, K. G. Bhattacharyya (2015). Removal of Ni (II) ions from aqueous solution by using low cost biosorbent prepared from jackfruit (Artocarpus heterophyllus) leaf powder. *Indian Journal of Chemical Technology,* 22, 322–327.

Brunauer, S., L. S. Deming, W. E. Deming, E. Teller (1940). On a theory of the van der Waals adsorption of gases. *Journal of the American Chemical Society,* 62, 1723–1732. https://doi.org/10.1021/ja01864a025

Cantuaria, M. L., A. F. de Almeida Neto, E. S. Nascimento, M. G. Vieira (2016). Adsorption of silver from aqueous solution onto pre-treated bentonite clay: Complete batch system evaluation. *Journal of Cleaner Production,* 112, 1112–1121.

Castañeda-Figueredo, J. S., A. I. Torralba-Dotor, C. C. Pérez-Rodríguez, A. M. Moreno-Bedoya, C. S. Mosquera-Vivas (2022). Removal of lead and chromium from solution by organic peels: Effect of particle size and bio-adsorbent. *Heliyon,* 8, e10275. https://doi.org/10.1016/j.heliyon.2022.e10275

Chakrabortya, R., A. Asthanaa, A. K. Singha, B. Jaina, A. B. Susanb (2022). Adsorption of heavy metal ions by various low-cost adsorbents: A review. *International Journal of Environmental Analytical Chemistry*, 102(2), 342–379. https://doi.org/10.1080/03067319.2020.1722811

Chakraborty, R., A. Asthana, A. K. Singh, B. Jain, A. B. H. Susan (2020). Adsorption of heavy metal ions by various low-cost adsorbents: A review. *International Journal of Environmental Analytical Chemistry*, 342–379. https://doi.org/10.1080/03067319.2020.1722811

Chen, X., M. F. Hossain, C. Duan, J. Lu, Y. F. Tsang, M. S. Islam, et al. (2022). Isotherm models for adsorption of heavy metals from water- A review. *Chemosphere*, 307, 135545. https://doi.org/10.1016/j.chemosphere.2022.135545

Chen, T., T., Da, Y. Ma (2021). Reasonable calculation of the thermodynamic parameters from adsorption equilibrium constant. *Journal of Molecular Liquids*, 322, 114980. https://doi.org/10.1016/j.molliq.2020.114980

Cornell, R. M., U. Schwertmann (1998). *The Iron Oxide*, 1st edition. VCH publishers.

Corrêa, R. A., L. A. Calçada, R. P. Peçanha (2007). Development of a fluidized bed system for adsorption of phenol from aqueous solutions with commercial macroporous resins. *Brazilian Journal of Chemical Engineering*, 24(1), 15–28.

Crawford, R. J., I. H. Harding, D. E. Mainwaring (1996). The zeta potential of iron and chromium hydrous oxides during adsorption and coprecipitation of aqueous heavy metals. *Jourma of Colloid and Interface Science*, 181, 561–570.

Cruz-Olivares, J., C. Pérez-Alonso, C. Barrera-Díaz, F. Ureña-Nuñez, M. Chaparro Mercado B. Bilyeu (2013). Modeling of lead (II) biosorption by residue of allspice in a fixed-bed column. *Chemical Engineering Journal*, 228, 21–27.

Darweesh M. A, M. Y. Elgendy, M. I. Ayad A. M. Mahmed, N. M. K. Elsayed, W. AHammad (2022). Sorption isotherm, kinetic, and optimization studies for copper (II) removal from aqueous solution systems by banana leaves and derived activated carbon. *South African Jounal of Chemical Engineering*, 40, 10–20. https://doi.org/10.1016/j.sajce.2022.01.002

Das, J., R. Saha, H. Nath, A. Mondal, S. Nag (2022). An eco-friendly removal of Cd(II) utilizing banana pseudo-fibre and Moringa bark as indigenous green adsorbent and modelling of 59 adsorption by artificial neural network. *Environmental Science and Pollution Research*, 29, 86528–86549. https://doi.org/10.1007/s11356-022-21702-z

Davis, J. A., J. O. Leckie (1978). Surface ionization and complexation at the oxide/water interface. II. Surface properties of amorphous iron oxyhydroxide and adsorption of metal ions. *Journal of Colloid and Interface Science*, 67, 90–107.

Dawood, S. (2018). Synthesis and characterization of biomass and clay minerals based adsorbents for the removal of cationic dye and metal ion from wastewater by adsorption. *Ph.D. thesis*.2018, Curtin University Liberary, Bentley, 6845, AUSTRALI. URI. https://hdl.handle.net/20.500.11937/69346

Dawood, S., T. K. Sen, C. Phan (2017). Synthesis and Characterisation of Pine Cone bio-char in the removal of organic and inorganic pollutants from aqueous solution by adsorption: Kinetic, equilibrium, mechanism and thermodynamic. *Bioresource and Biotechnology*, 246, 76–78. https://doi.org/10.1016/j.biortech.2017.07.019

Dawood, S. T., K. Sen (2012). Removal of anionic dye Congo red from aqueous solution by raw pine and acid-treated pinecone powder as adsorbent: Equilibrium, thermodynamic, kinetics, mechanism and process design. *Water Research*, 46, 1933–1946. https://doi.org/10.1016/j.watres.2012.01.009

Dawood, S., T. K. Sen (2012). Author's responses to the comment by Canzano Et Al and also Corrigendum to removal of anionic dye Congo Red from aqueous solution by raw Pine and acid-treated Pine Cone Powder as Adsorbent: Equilibrium, thermodynamic, kinetics, mechanism and process design. *Water Research*, 46, 4316–4317. https://doi.org/10.1016/j. watres.2012.05.043

Dawood, S., T. K. Sen, C. Phan (2014). Preparation and characterization of novel activated carbon from waste biomass pine cone and its application in Congo red dye adsorption from aqueous solution: Equilibrium, thermodynamic, kinetics, mechanism and process design. *Water, Air & Soil Pollution*, 225, 1818. https://doi.org/10.1007/s11270-013-1818-4

Dawood, S., T. K. Sen, C. Pan (2018). Performance and dynamic modelling of biochar and kaolin packed bed adsorption column for aqueous phase methylene blue (MB) dye removal. *Environmental Technology*, 40(25), 1–35. https://doi.org/10.1080/09593330.2018.1491065

Dong, Y., H. Lin (2017). Competitive adsorption of Pb(II) and Zn(II) from aqueous solution by modified beer lees in a fixed bed column. *Process Safety and Environmental Protection*, 111, 263–269. https://doi.org/10.1016/j.psep.2017.06.016

Dutta, B. K. (2007). *Principle of Mass Transfer and Separation Process. PHI learning, India*, 1st edition, ISBN-10: 8120329902, ISBN-13: 978–8120329904

Dzombak, D. A., F. M. M. Morel (1990). *Surface Complexing Modelling. Hydrous ferric Oxide*, John Wiley, 393–393.

Dzombak, D., F. M. M. Morel (1988). Sorption of cadmium on hydrous ferric oxide at high sorbate/sorbent ratios: Equilibrium, kinetics, and modelling. *Journal of Colloid and Interfact Science*, 112, 589–597.

El-Baz, A. A., I. A. Hendyb, A. M. Dohdohc, M. I. Soru (2020). Adsorption technique for pollutants removal; current new trends and future challenges – A Review. *The Egyptian International Journal of Engineering Sciences and Technology*, 32, 1–24. https://eijest.journals.ekb.eg

Fan, L. T., Y. C. Yang, C. Y. Wen (1959). Semifluidization: Mass transfer in semi fluidized beds. *AIChE Journal*, 5, 407.

Fan, L., C. Y. Wen (1961). Mechanics of semifluidization of single size particles in solid-liquid systems. *AIChE Journal*, 7(4), 606–610.

Faust, S. D., O. M. Aly (2013). *Adsorption Processes for Water Treatment*. Elsevier.

Francisco, A., T. K. Sen (2009). Removal of zinc metal ion (Zn^{2+}) from its aqueous solution by kaolin clay minerals: A kinetic and equilibrium study. *Colloids and Surfaces A*. 348, 100–108.

Freundlich, H. M. F (1906). Ober dies adsorption in losungen. *Zeitschrift Physikalische Chemie*, 57, 385e470.

Febrianto, J., A. N. Kosasih, J. Sunarso, Y.-H. Ju, N. Indraswati, S. Ismadji (2009). Equilibrium and kinetic studies in adsorption of heavy metals using biosorbent: A summary of recent studies. *Journal of Hazardous Materials*, 162(2), 616–645.

Geankoplis, C. J. (2003). *Transport Processes and Separation Process Principles (Includes Unit Operation)*. 4th edition. Prentice Hall. ISBN: 013101367X

Ghaedi, M., A. Najibi, H. Hossainian, A. Shokrollahi, M. Soylak (2012). Kinetic and equilibrium study of Alizarin Red S removal by activated carbon. *Toxicological & Environmental Chemistry*, 94(1), 40–48.

Ghazy, A. R., K. El-Refaie, A. Hamed, A. Abosharaf, M. E. Kenawy, M. Bishr, M. M. Azaam (2024). Poly acrylonitrile Schiff-bases nanofibers: Antibacterial and dye removal potency. *Journal of Water Process Engineering*, 64, 105645. https://doi.org/10.1016/j.jwpe.2024.105645

Ghosh, N., S. Sen · Goutam Biswas · Atul Saxena, P. K. Haldar (2023). Adsorption and desorption study of reusable magnetic iron Oxide nanoparticles modifed with justicia adhatoda leaf extract for the removal of textile dye and antibiotic. *Water Air Soil Pollution*, 234, 202. https://doi.org/10.1007/s11270-023-06217-8

Gul, Salma, S. Gul, H. Gul, N. Raza, A. Azzouz, M. R. Elamin, L. Khezami (2024). Enhanced adsorptive removal of malachite green in environmental samples using Java plum leaves: From equilibrium to mechanism studies. *Biomass Conversion and Biorefnery*, 15, 4225–4241. https://doi.org/10.1007/s13399-024-05567-5

Ho, Y.-S (2006). Review of second-order models for sorption systems. *Journal of Hazardous Materials*, B136, 681–689. https://doi.org/10.1016/j.jhazmat.2005.12.043

Ho, Y.-S., G. McKay (1999). Pseudo-second order model for sorption processes. *Process Biochemistry*, 34(5), 451–465.

Hohl, H., W. Stumm (1976). Interaction of Pb^{2+} with hydrous Al2O3. *Journal of Colloid Interface Science*, 55, 281–288.

Huang, Y., W. Su, R. Wang, T. Zhuang (2019). Removal of typical industrial gaseous pollutants: From carbon, zeolite, and metal-organic frameworks to molecularly imprinted adsorbent. *Aerosol and Air Quality Research*, 19, 2130–2150. https://doi.org/10.4209/aaqr.2019.04.0215

Hutchins, R. (1973). New method simplifies design of activated-carbon systems. *Chemical Engineering*, 80(19), 133–138.

Jena, H. M., G. K. Roy, B. C. Meikap (2009). Hydrodynamics of regular particles in a liquid-solid semi-fluidized bed. *Powder Technology*, 196(3), 246–256.

Kamal, I. (2020). Batch and Continuous Adsorption Processes for Removing Contaminants from Wastewater. *Conference: 2020 International Conference on Hydrology and Water Resources Engineering (HWRE 2020)* 8-10 May 2020. https://doi.org/10.13140/RG.2.2.34695.62885

Kapur, M., M. K. Mondal (2016). Design and model parameters estimation for fixed–bed column adsorption of Cu (II) and Ni (II) ions using magnetized saw dust. *Desalination and Water Treatment*, 57(26), 12192–12203.

Katz, L. E., K. F. Hayes (1995). Surface complexation modelling 1. Strategy for modelling monomer complex formation at moderate surface coverage. *Journal of Colloid and Interface Science*, 170, 477–490.

Kaur, S., S. Rani, R. K. Mahajan (2013). Adsorption kinetics for the removal of Hazardous dye Congo red by Biowaste materials as adsorbents. *Journal of Chemistry*, 12, Article ID 628582. https://doi.org/10.1155/2013/628582

Konicki, W., D. Sibera, E. Mijowska, Z. Lendzion-Bieluń, U. Narkiewicz (2013). Equilibrium and kinetic studies on acid dye Acid Red 88 adsorption by magnetic ZnFe2O4 spinel ferrite nanoparticles. *Journal of Colloid and Interface Science*, 398, 152–160.

Kovach, J. L (1997). Gas Phase Adsorption, in: P. A. Schweitzer (Ed). *Handbook of Separation Techniques for Chemical Engineers*, McGraw-Hill, New York, pp.3–47.

Kratochvil, D., B. Volesky (1998). Advances in the biosorption of heavy metals. *Trends in Biotechnology*, 16(7), 291–300.

Kulkarni, S. J., J. P. Kaware (2015). Analysis of packed bed adsorption column with low cost adsorbent for cadmium removal. *International Journal of Thermal & Environmental Engineering*, 9(1), 17–24.

Lagergren, S. (1898). About the theory of so called adsorption of soluble substances. Zur theoric der sogenannten adsorption geloster stoffe. *Kungliga Svenska Vetenskapsakad Handlinge*, 24(4), 1–39, Band 24.

Lallan S. Y., K. M. Bijay, K. Arvind (2017). Batch and column studies for phenol removal from aqueous solutions using laboratory prepared low-cost activated carbon as an adsorbent. *Chemical Technology Indian Journal*, 12(2), 114.

Langmuir, I. (1918). The adsorption of gases on plane surfaces of glass, mica and platinum. *Journal of the American Chemical Society*, 40(9), 1361–1403.

Lima J. Z, EF. da Silva, C. Patinha, N. Durães, E. M. Vieira, V. G. S. Rodrigues (2022). Sorption of arsenic by composts and biochars derived from the organic fraction of municipal solid wastes: Kinetic, isotherm and oral bioaccessibility study. *Environmental Research*, 204, 111988. https://doi.org/10.1016/j.envres.2021.111988

Liu, C., X. Yan, Y. Chen, Y. Zhou, H. Wu, H. Wang, H.-X. Zhang, J. Yang, K.-B. Yoon (2024). Biochars and modified-biochars for toxic-metal/metalloid ions sorption in various mixed solution systems: A review on kinetic and isotherm models. *Desalination and Water Treatment*, 319, 100404. https://doi.org/10.1016/j.dwt.2024.100404

Liu, C., X. Yan, Y. J. Chen, Y. Y. Zhou, H. Wu, H. Wang et al. (2023). N-cyanoguanidine modified black peanut shell biochar: Fabrication and its sorption for Cu (II) and Co(II) in a single and mixed solutions. *Mater Today Sustainability*, 24, 100540. https://doi.org/10.1016/j.mtsust.2023.100540

Long, Y., D. Lei, J. Ni, Z. Ren, C. Chen, H. Xu (2014). Packed bed column studies on lead (II) removal from industrial wastewater by modified Agaricus bisporus. *Bioresource Technology*, 152, 457–463.

Madala, S., V. N. R. Mudumala, S. Vudagandla, K. Abburi (2015). Modified leaf biomass for Pb (II) removal from aqueous solution: Application of response surface methodology. *Ecological Engineering*, 83, 218–226.

McKinley, J. P., E. A. Jenne (1991). Experimental investigation and review of the 'Solid concentration effect in adsorption studies. *Environmental Science and Technology*, 25, 2082–2087.

Mendizabal, E., N. Ríos-Donato, I. Paulina, I. Verduzco-Navarro (2023). Removal of arsenate by fixed-bed columns using chitosan-magnetite hydrogel beads and chitosan hydrogel beads: Effect of the operating conditions on column efficiency. *Gels*, 9, 825. https://doi.org/10.3390/ gels9100825

Milonjic, S. K (2007). A consideration of the correct calculation of thermodynamic parameters of adsorption. *Journal of the Serbian Chemical Society*, 72(12), 1363–1367. https://doi.org/10.2298/JSC0712363M

Miraboutalebi, S. M., S. K. Nikouzad, M. Peydayesh, N. Allahgholi, L. Vafajoo, G. McKay (2017). Methylene blue adsorption via maize silk powder: Kinetic, equilibrium, thermodynamic studies and residual error analysis. *Process Safety and Environmental Protection*, 106, 191–202.

Montgomery, D. C, E. A Peck, G. G Vining (2012). *Introduction to Linear Regression Analysis*, 5th Edition. Wiley.

Najafi, H., E. Pajootan, A. Ebrahimi, M. Arami (2015). The potential application of tomato seeds as low-cost industrial waste in the adsorption of organic dye molecules from colored effluents. *Desalination and Water Treatment*, 57(32), 15026–15036, 2016. https://doi.org/10.1080/19443994.2015.1072060.

Nawaz, S., H. N. Bhatti, T. H. Bokhari, S. Sadaf (2014). Removal of Novacron Golden Yellow dye from aqueous solutions by low-cost agricultural waste: Batch and fixed bed study. *Chemistry and Ecology*, 30(1), 52–65.

Nwabanne, J., P. Igbokwe (2012). Adsorption performance of packed bed column for the removal of lead (II) using oil palm fibre. *International Journal of Applied Science and Technology*, 2(5), 106–115.

Oladoja, N., C. Aboluwoye, Y. Oladimeji, A. Ashogbon, I. Otemuyiwa (2008). Studies on castor seed shell as a sorbent in basic dye contaminated wastewater remediation. *Desalination*, 227(1–3), 190–203.

Özkaya, B., A. H. Kaksonen, E. Sahinkaya, J. A. Puhakka (2019). Fluidized bed bioreactor for multiple environmental engineering solutions. *Water Research*, 150, 452–465.

Papageorgiou, K., F. K. Katsaros, E. P. Kouvelos, N. K. Kanellopoulos (2009). Prediction of binary adsorption isotherms of Cu2+, Cd2+ and Pb2+ on calcium alginate beads from single adsorption data. *Journal of Hazardous Materials*, 162, 1347–1354. https://doi.org/10.1016/j.jhazmat.2008.06.022

Patel, H. (2019). Fixed bed column adsorption study: A comprehensive review. *Applied Water* 8 Science, 9(45), 1–17.

Patel, H. (2021). Comparison of batch and fixed bed column adsorption: A critical review. *International Journal of Environmental Science and Technology*, 19, 10409–10426.

Radhika R., T. Jayalatha, G. Rekha Krishnan, J. Salu, R., Rajeev B. K. George (2020). Adsorption performance of packed bed column for the removal of perchlorate using modified activated carbon. *Process Safety and Environmental Protection*, 117, (2018) 350–362. https://doi.org/10.1016/j.psep.2018.04.026

Sahu, C., F. Khan, P. K. Pandey, M. Pandey (2017). Biosorptive removal of toxic contaminant lead from wastewater. *Asian Journal of Chemistry*, 29(3), 650–656.

Sakhiya, A. K., V. K. Vijay, P. Kaushal (2022). Efficacy of rice straw derived biochar for removal of Pb+2 and Zn+2 from aqueous: Sorption, thermodynamic and cost analysis. *Bioresource Technology Reports*, 17, 100920. https://doi.org/10.1016/j.biteb.2021.10092

Samanth, A., R. Selvaraj, G. Murugesan, T. Varadavenkatesan, R. Vinayagam (2024). Efficient adsorptive removal of 2,4-dichlorophenoxyacetic acid (2,4-D) using biomass derived magnetic activated carbon nanocomposite in synthetic and simulated agricultural runoff water. *Chemosphere*, 361, 142513. https://doi.org/10.1016/j.chemosphere.2024.142513

Sazali, N., Z. Harun, N. Sazali (2020). A review on batch and column adsorption of various adsorbent towards the removal of heavy metal. *Journal of Advanced Research in Fluid Mechanics and Thermal Sciences*, 67(2), 66–88. 22.

Schindler, P. W., H. R. Kamber (1968). Die aciditat von silanolgruppen. *Helvetica Chimica Acta*, 51, 1781–1786.

Schubert, M., Larachi et al. (2009). Phase holdups in three-phase semifluidized beds and the generalized bubble wake model. *Industrial and Engineering Chemistry Research*, 48(18), 8393–8401.

Seader, J. D., E. J. Henley (1998). *Separation Process Principles*. Wiley.

Sen, T. K. (2001). Ph. D. thesis entitled Studies on Colloid-associated contaminant transport in saturated porous media at Indian Institute of Technology, Bombay Library, Powai, Maharashtra, India.

Sen, T. K. (2012). A simple surface-complex model (SCM) for metal ion adsorption mechanism. *World Academy of Science, Engineering and Technology*, 66, 1367–1371. https://hdl.handle.net/20.500.11937/10538

Sen, T. K. (2023a). Application of synthesized biomass bamboo charcoal–iron oxide "BC/Fe" nanocomposite adsorbents in the removal of cationic methylene blue dye contaminants from wastewater by adsorption. *Sustainability*, 15, 8841. https://doi.org/10.3390/su15118841

Sen, T. K. (2023b). Adsorptive removal of dye (methylene blue) organic pollutant from water by pine tree leaves biomass adsorbent. *Processes*, 2023, 11, 1877. https://doi.org/10.3390/pr11071877

Sen, T. K., S. P. Mahajan, K. C. Khilar (2002). Adsorption of Cu+2 and Ni+2 on iron oxide and kaolin and its importance on Ni2+ transport in porous media. *Colloids and Surfaces A*, 211, 91–102.

Shan, R., Y. Shi, J. Gu, Y. Wang, H. Yuan (2020). Single and competitive sorption affinity of heavy metals toward peanut shell-derived biochar and its mechanisms in aqueous systems. *Chinese Journal of Chemical Engineering*, 28, 1375–1383. https://doi.org/10.1016/j.cjche.2020.02.012

Sivalingam, S., A. Sowmiya (2024). Industrial biomass waste as an economical, potential adsorbent for removing the Bismarck Brown R dye and zinc metal ions from effluents. *Environmental Science Advances*, 3, 732–750. https://doi.org/10.1039/d3va00314k

Sivalingam, S., V. Gopal (2024). Low-cost adsorbent from biomass for removal of Fe(II) and Mn(II) for water treatment: Batch and column adsorption study. *Chemical Papers*, https://doi.org/10.1007/s11696-024-03438-x

Soetaredjo, F. E., A. Kurniawan, L. Ong, D. R. Widagdyo. S. Ismadji (2014). Investigation of the continuous flow sorption of heavy metals in a biomass-packed column: Revisiting the Thomas design model for correlation of binary component systems. *RSC Advances*, 4(95), 52856–52870.

Tan, K. L., B.H. Hameed (2017). Insight into the adsorption kinetics models for the removal of contaminants from aqueous solutions. *Journal of the Taiwan Institute of Chemical Engineers*, 74, 25–48. https://doi.org/10.1016/j.jtice.2017.01.024

Tan, T. C. N., T. K. Sen (2020). Author's responses to the comment by Jean-Claude Bollinger and also corrigendum to our recent article published in Separation Science and Technology online March 4, 2019. *Separation Science Technology*, 55(5), 825–827. https://doi.org/10.1080/01496395.2019.1706579

Tan, T. C. N., T. K. Sen (2019–2020). Aqueous-phase methylene blue (MB) dye removal by mixture of eucalyptus bark (EB) biomass and kaolin clay (KC) adsorbents: Kinetics, thermodynamics, and isotherm modelin. *Separation Science and Technology*, online March 4, 2019. *Separation science Technology*, 55(6), 1036–1050. https://doi.org/10.1080/01496395.2019.1580734

Thomas, H. C. (1944). Heterogeneous ion exchange in a flowing system. *Journal of the American Chemical Society*, 66(10), 1664–1666.

Tian, Y., X. Sun, N. Chen, X. Cui, H. Yu, Y. Feng, D. Xing, W. He (2024). Efficient removal of hexavalent chromium from wastewater using a novel sodium alginate-biochar composite adsorbent. *Journal of Water Process Engineering*, 64, 105655. https://doi.org/10.1016/j.jwpe.2024.105655

Tien, C. (1994). *Adsorption Calculation and Modelling*, 1st edition. Butterworth-Heinemann, ISBN 0-7506-9121-2.

Umembamalu, C. J., C. A. Igwegbe, E. U. Osuagwu, J. T. Nwabanne (2020). Packed bed column adsorption of oil and grease from refinery desalter effluent, using rice husks derived carbon as the adsorbent: Influence of process parameters and Bohart–Adams kinetics study. *World News of Natural Sciences*, 31, 155–174. www.worldnewsnaturalsciences.com

Vaghetti J. C. P, E. C. Lima, B. Royer, B. M. da Cunha, N. F. Cardoso, J. L. Brasil et al (2009). Pecan nutshell as biosorbent to remove Cu(II), Mn(II) and Pb(II) from aqueous solution systems. *Journal of Hazardous Materials*, 162, 270–280. https://doi.org/10.1016/j.jhazmat.2008.05.039

Vassileva, P., I. Uzunov, A. Detcheva, U. Snezhanka, D. Voykova (2019). Multi-component adsorption of heavy metal ions from aqueous solutions onto low-cost adsorbent based on rice Husks. *Integrative Sciences Journal Mechanization in Agriculture and Conserving of the Resources*, 181(5), 178–181.

Veeraraghavan, S., L. T. Fan, A. P. Mathews (1989). Modeling adsorption in liquid—Solid Fluidized beds. *Chemical Engineering Science*, 44(10), 2333–2345.

Vilvanathan, S., S. Shanthakumar (2017). Column adsorption studies on nickel and cobalt removal from aqueous solution using native and biochar form of Tectona grandis. *Environmental Progress & Sustainable Energy*, AIChE. 36(4), 1030–1038. https://doi.org/10.1002/ep.12567

Weber, W., J. Morris (1963). Kinetics of adsorption on carbon from solution. *Journal of the Sanitary Engineering Division, Proceeding of the American Society of Civil Engineers*, 89(17), 31–60.

Wen, X., Q. Du, H. Tang (1998). Surface complexation model for the heavy metal adsorption on natural sediment. *Environmental Science and Technology*, 32, 870–875.

Xu, Z., J.-G. Cai, B.-C. Pan (2013). Mathematically modelling fixed-bed adsorption in aqueous systems. *Journal of Zhejiang University-Science A (Applied Physics & Engineering)*, 14, 155–176.

Yaari, A. L., M. A. T. A. Saleh, O. Saber (2021). Removal of mercury from polluted water by a novel composite of polymer carbon nanofiber: Kinetic, isotherm and thermodynamic studies. *RSC Advances*, 11, 380–389.

Yaari, A. L. M., T. A. Saleh (2023). Removal of lead from wastewater using synthesized polyethyleneimine-grafted graphene oxide. *Nanomaterials*, 13, 1078. https://doi.org/10.3390/nano13061078

Yaghoobi-Khankhajeh, S., R. Alizadeh, R. Zarghami (2018). Adsorption modeling of CO2 in Fluidized Bed Reactor. *Chemical Engineering Research and Design*, 129, 111–121.

Yagub, M. T., T. K. Sen, S. Afroze, H. Ang (2014a). Dye and its removal from aqueous solution by adsorption: A review. *Advances in Colloid and Interface Science*, 209, 172–184.

Yagub, M. T., T. K. Sen, S. Afroze, H. M. Ang (2014b). Fixed-bed dynamic column adsorption study of methylene blue (MB) onto pine cone. *Desalination and Water Treatment*, 55(4), 1026–1039, 2015. https://doi.org/10.1080/19443994.2014.924034

Yaneva, Z. L., N. V. Georgieva (2012). Insights into congo red adsorption on agro-industrial materials- spectral, equilibrium, kinetic, thermodynamic, dynamic and desorption studies. A Review. *International Review of Chemical Engineering*, 4(2), 127–146.

Yao, Y., H. Zuo, Y. Liu, S. Pang, L. Lan, F. Yao, Y. Wuab, Z. Li (2024). Efficient dye adsorption of mesoporous activated carbon from bamboo parenchyma cells by phosphoric acid activation. *RSC Advances*, 14, 1287.

Yi, H., Z. Wang, H. Liu, X. Tang, D. Ma, S. Zhao, B. Zhang,F. Gao, Y. Zuo (2014). Adsorption of SO 2, NO, and CO 2 on activated carbons: Equilibriumand thermodynamics. *Journal of Chemical & Engineering Data*, 59(5), 1556–1563. https://doi.org/10.1021/je4011135

Yildiz, H. A. (2024). The production of a novel adsorbent from forest waste (Platanus orientalis L.) for dye adsorption: Adsorption process optimization and experimental design. *Materials Science & Engineering B*, 304, 117366. https://doi.org/10.1016/j.mseb.2024.117366

Yoon, Y. H., J. H. Nelson (1984). Application of gas adsorption kinetics I. A theoretical model for respirator cartridge service life. *American Industrial Hygiene Association Journal*, 45, 509–516.

Zendehdel, M., H. Mohammadi (2018). The fixed-bed column study for heavy metals removal from the wastewater by poly acrylamide co-acrylic acid/clinoptilolite nanocomposite. *Nanoscience and Nanotechnology-Asia*, 8(1), 67–74.

Zhang, Y., M. Li, J. Li, Y. Yang, X. Liu (2019). Surface modified leaves with high efficiency for the removal of aqueous Cr(VI). *Applied Surface Science*, 484, 189–196. https://doi.org/10.1016/j.apsusc.2019.04.088

Zhang, S., Z. Li, Y. Yang, W. Jian, D. Ma, F. Jia (2020). Kinetics and thermodynamics of SO2 adsorption on metal-loaded multiwalled carbon nanotubes. *Open Physics*, 18, 1201–1214. https://doi.org/10.1515/phys-2020-0215

Zhang, X., H. Xu, M. Xi, Z. Jiang (2023). Removal/adsorption mechanisms of Cr(VI) and natural organic matter by nanoscale zero-valent iron-loaded biochar in their coexisting system. *Journal of Environmental Chemical Engineering*, 11, 109860. https://doi.org/10.1016/j. jece.2023.109860

Zhao, Z., B. Wang, B. K. G. Theng, X. Lee, X. Zhang, M. Chen, P. Xu (2022). Removal performance, mechanisms and influencing factors of biochar for air pollutants: A critical review. *Biochar*, 4, 30. https://doi.org/10.1007/s42773-022-00156-z

Zhou, X., H. Yi, X. Tang, H. Deng, H. Liu (2012). Thermodynamics for the adsorption of SO_2, NO and CO_2 from flue gas on activated carbon fiber. *Chemical Engineering Journal*, 200–202, 399–404. https://doi.org/10.1016/j.cej.2012.06.013

Index

E

F

G

H

X

Y

Z

For Product Safety Concerns and Information please contact our EU
representative GPSR@taylorandfrancis.com
Taylor & Francis Verlag GmbH, Kaufingerstraße 24, 80331 München, Germany

www.ingramcontent.com/pod-product-compliance
Ingram Content Group UK Ltd.
Pitfield, Milton Keynes, MK11 3LW, UK
UKHW020204200926
475702UK00004B/603

* 9 7 8 1 0 3 2 4 1 9 7 7 0 *